Alchemy in Action:
Metallurgy, Refractory Materials, and the Science of Transformation

RAHUL KUMAR

ISBN: 9798394989261
Imprint: Independently published

DEDICATION

This book is dedicated to my mother Mrs. Gaytri Prasad and my father
Mr. Ranjan Kumar

CONTENTS

Industry Practices

ACKNOWLEDGEMENT

I would like to express my heartfelt gratitude to all those who have contributed to the creation and success of "Alchemy in Action: Metallurgy, Refractory Materials, and the Science of Transformation."
This book would not have been possible without the support, guidance, and encouragement of numerous individuals who have played a significant role in my journey.

First and foremost, I extend my deepest appreciation to my family, my parents, and my sisters. Their unwavering belief in me, their constant encouragement, and their sacrifices have been the bedrock of my achievements. Their love and support have given me the strength and motivation to pursue my passions.

I am immensely grateful to my mentors, supervisors, and colleagues at Jindal Stainless Limited and Calderys India Refractories Limited. Their valuable insights, guidance, and expertise have shaped my understanding of the metallurgy and refractory materials field. Their unwavering support and willingness to share their knowledge have been instrumental in my professional growth.

I extend my sincere appreciation to the experts and researchers in the field of metallurgy and refractory materials who have contributed to the body of knowledge in this area. Their groundbreaking work and research findings have been the foundation upon which this book is built.

Special thanks to my senior at Calderys India Refractories Limited, Mr. Shamim Ahmed for motivating me to write this book.

To my friends and colleagues who have provided valuable feedback, assistance, and encouragement along the way, I am deeply thankful. Your support has been invaluable, and I am grateful for the enriching discussions and exchange of ideas.

Last but not least, I want to express my profound gratitude to the readers of this book. Your interest in the subject matter and your willingness to explore the world of metallurgy and refractory materials inspire me to continue sharing knowledge and contributing to the advancement of our industry.

In conclusion, this book is the result of the collective efforts and support of numerous individuals, and I am immensely grateful for each and every one of you. Your contributions have made "Alchemy in Action" a reality, and I hope it serves as a valuable resource and inspiration for all those who seek to delve into the fascinating world of metallurgy and refractory materials.

Thank you.

Rahul Kumar

Dhanbad, May 2023.

ABOUT THE AUTHOR

Rahul Kumar is a passionate metallurgical engineer, author, and industry professional dedicated to exploring the fascinating world of metals and their transformations. With a diverse background in refractory materials, caster operations, and mould flux development, he brings a wealth of practical experience and expertise to the field of metallurgy.

After graduating from BIT Sindri, having spent several years in the steel industry, Rahul Kumar has had the opportunity to work with renowned organizations such as Jindal Stainless Limited and Calderys India Refractories Limited. These experiences have allowed him to witness firsthand the intricacies of metallurgical processes and the critical role of refractory materials in shaping the industry.

Driven by an insatiable curiosity and a desire to share knowledge, Rahul Kumar embarked on a journey to create "Alchemy in Action: Metallurgy, Refractory Materials, and the Science of Transformation." This technical book is the culmination of their expertise, research, and passion for the subject.

Throughout his career, Rahul Kumar has not only excelled in his professional endeavours but has also pursued personal interests in writing and creative expression. He is a proficient writer of Hindi poems and occasional short stories, finding joy and inspiration in the written word.

Beyond his technical expertise, Rahul Kumar is a firm believer in the power of education and lifelong learning. He has dedicated himself to continuous growth, for personal and professional development.

Originally hailing from the vibrant city of Dhanbad, Jharkhand, Rahul Kumar brings a rich cultural heritage and a deep appreciation for diversity to his work. His upbringing in a close-knit community has instilled in them values of perseverance, dedication, and a strong work ethic.

With the release of "Alchemy in Action," Rahul Kumar aspires to inspire and empower professionals, students, and enthusiasts in the field of metallurgy. Through his comprehensive exploration of metallurgical transformations, he hopes to ignite a sense of curiosity and open new avenues of understanding for readers worldwide.

Connect with Rahul Kumar on LinkedIn and stay updated with his latest work, insights, and upcoming projects. Join them on their journey of unravelling the mysteries and unlocking the potential of metallurgical alchemy.

LinkedIn: **https://www.linkedin.com/in/rahul-metallurgist**

Email: **rahul1rockson@gmail.com**

Follow Rahul Kumar on social media for the latest updates:

Twitter: **https://twitter.com/rahulmetal5**

Instagram: **https://instagram.com/so_called_kumar**

Facebook: **https://www.facebook.com/profile.php?id=100004468001786**

For inquiries, speaking engagements, or collaborations, please reach out using the provided contact information.

Thank you for your support and for joining Rahul Kumar on his captivating exploration of metallurgy, refractory materials, and the science of transformation.

FROM AUTHOR'S DESK

Dear Readers,

It is with great pleasure and excitement that I present to you "Alchemy in Action: Metallurgy, Refractory Materials, and the Science of Transformation." As the author of this book, I am honoured to share my passion and knowledge in the field of metallurgy with you.

From a young age, I have been captivated by the transformative power of metals and the intricate processes that govern their behaviour. This fascination led me to pursue a career as a metallurgical engineer, where I have had the privilege of working in various capacities within the steel industry.

Through my experiences at Jindal Stainless Limited and Calderys India Refractories Limited, I have witnessed firsthand the remarkable impact that metallurgy and refractory materials have on the creation of extraordinary alloys. The artistry and science behind transforming raw materials into durable and versatile metals have always intrigued me, motivating me to delve deeper into the subject.

"Alchemy in Action" is the result of years of study, research, and practical experience in the field. In this book, I aim to unravel the mysteries of metallurgical transformations and shed light on the essential principles and processes that underpin them. It is my sincerest hope that this comprehensive guide will serve as a valuable resource for professionals, students, and enthusiasts alike, offering insights and practical knowledge that can be applied in real-world scenarios.

Throughout the pages of this book, you will explore the scope of metallurgy, delve into the nuances of refractory materials, and gain a deeper understanding of casting processes. I have endeavored to strike a balance between technical depth and accessibility, presenting the material in a manner that engages readers from various backgrounds and levels of expertise.

It is my belief that knowledge is most meaningful when shared, and I am grateful for the opportunity to contribute to the body of knowledge in the field of metallurgy. By sharing my experiences, research, and insights, I hope to inspire others to explore the boundless possibilities and transformative potential that lie within the realm of metallurgical science.

I would like to express my deepest gratitude to all those who have supported me throughout this journey. To my family, mentors, colleagues, and friends who have believed in me and provided unwavering support, thank you for your encouragement and guidance. Your belief in my abilities has been the driving force behind the completion of this book.

To the readers of "Alchemy in Action," I hope that this book ignites your curiosity, broadens your understanding, and empowers you to embark on your own transformative journeys within the realm of metallurgy. I invite you to join me in unlocking the secrets, unraveling the complexities, and embracing the enchanting world of metallurgical alchemy.

Thank you for embarking on this journey with me.

Sincerely,

Rahul Kumar

Chapter 1 | The Future of Metallurgy: Trends and Predictions

Metallurgy has been a cornerstone of human civilization for thousands of years. From the Bronze Age to the Space Age, metallurgy has played a critical role in shaping the world as we know it. Today, metallurgy is more important than ever, with applications in everything from transportation and energy production to medicine and electronics. In this chapter, we will explore the future of metallurgy, including the trends and predictions that are shaping this vital field.

The Role of Advanced Materials in Metallurgy

One of the most significant trends in metallurgy is the development of advanced materials. Researchers and engineers are constantly exploring new ways to create materials with enhanced properties, such as greater strength, durability, and resistance to corrosion. Advances in material science have already led to the creation of high-strength steels, titanium alloys, and composite materials. In the future, we can expect to see even more advanced materials that push the boundaries of what is possible in metallurgy.

The Rise of Additive Manufacturing

Another trend that is transforming the world of metallurgy is the rise of additive manufacturing, or 3D printing. Additive manufacturing allows engineers to create complex metal parts and structures with incredible precision and accuracy. This technology is already being used in a variety of industries, including aerospace, medical devices, and automotive manufacturing. As the technology continues to advance, we can expect to see even more applications for additive manufacturing in metallurgy.

The Importance of Sustainability

Sustainability is becoming an increasingly important consideration in metallurgy. As the demand for metals and materials continues to grow, there is a need to find ways to reduce waste and minimize the environmental impact of metallurgical processes. One trend that is gaining traction is the use of renewable energy sources to power metallurgical operations. In addition, there is a growing focus on recycling and reusing metals to minimize the need for new mining and extraction.

The Role of Digitalization and Big Data

Digitalization and big data are also having a significant impact on metallurgy. By collecting and analyzing large amounts of data, engineers and researchers can gain new insights into metallurgical processes and optimize their performance. Digitalization is also enabling greater automation in metallurgical operations, improving efficiency and reducing the risk of human error.

The Future of Metallurgy: Predictions

So what does the future hold for metallurgy? Here are a few predictions:

1. Advances in materials science will continue to drive innovation in metallurgy, leading to the development of new alloys and composites with even greater properties.

2. Additive manufacturing will become more widespread, enabling engineers to create complex metal parts and structures with greater ease and precision.

3. Sustainability will become an increasingly important consideration in metallurgy, with a focus on reducing waste, minimizing environmental impact, and promoting recycling and reuse.

4. Digitalization and big data will continue to transform metallurgical

operations, enabling greater automation, efficiency, and optimization.

Conclusion

Metallurgy is a dynamic field that is constantly evolving. With advances in materials science, additive manufacturing, sustainability, and digitalization, the future of metallurgy is bright. As we look ahead, it is clear that metallurgy will continue to play a critical role in shaping the world we live in, from the smallest medical devices to the largest infrastructure projects.

Chapter 2 | A Beginner's Guide to Metallurgy: Key Concepts and Definitions

Metallurgy is the study of metals and their properties, including their physical, chemical, and mechanical behaviour. It is a branch of materials science that deals with the processing, properties, and performance of metals and alloys.

This chapter serves as a beginner's guide to metallurgy, providing key concepts and definitions that are essential to understanding the field. It will cover the basics of metallurgy, including the properties of metals and alloys, the process of extracting and refining metals, and the different types of metallurgical processes.

Properties of Metals and Alloys

Metals are characterized by their metallic luster, good thermal and electrical conductivity, high melting and boiling points, and malleability. Alloys, on the other hand, are mixtures of two or more metals or a metal and a non-metal. Alloys exhibit properties that are different from those of their component elements, making them more useful for specific applications.

Some of the most important properties of metals and alloys include:

1. Density: Metals and alloys are usually dense materials, with densities ranging from 2 to 22 g/cm3.

2. Strength: Metals and alloys are generally strong and stiff, with high tensile and compressive strengths.

3. Ductility: Most metals and alloys are ductile, meaning they can be drawn into wires or rolled into sheets without breaking.

4. Hardness: The hardness of a metal or alloy determines its resistance to indentation or scratching.

5. Corrosion resistance: Some metals and alloys are highly resistant to corrosion, while others are more susceptible to oxidation or other forms of degradation.

Extracting and Refining Metals

The process of extracting metals from ores and refining them involves several stages, including mining, crushing, grinding, smelting, and refining. The goal of this process is to separate the desired metal or metals from the other materials in the ore and purify it to the required level of purity.

The most common techniques used for extracting and refining metals include:

1. Pyrometallurgy: Involves the use of heat to extract and refine metals, usually by smelting or roasting the ore.

2. Hydrometallurgy: Involves the use of liquid solvents to extract metals from ores or concentrates.

3. Electrometallurgy: Involves the use of electricity to extract and refine metals, such as in the production of aluminium.

Metallurgical Processes

Metallurgical processes are used to transform raw materials into finished products. Some of the most common metallurgical processes include:

1. Casting: Involves pouring molten metal into a mold to create a specific shape.

2. Rolling: Involves passing metal through a series of rollers to reduce its thickness and create a uniform shape.

3. Forging: Involves heating metal and then shaping it by hammering, pressing, or rolling.

4. Welding: Involves joining two or more pieces of metal together using heat or pressure.

5. Heat Treatment: Involves heating and cooling metal to alter its properties, such as its hardness or strength.

Conclusion

Metallurgy is a fascinating field that has a wide range of applications in various industries, including manufacturing, construction, aerospace, and medicine. This chapter provided a basic overview of metallurgy, including the properties of metals and alloys, the process of extracting and refining metals, and the different types of metallurgical processes. It is important to have a solid understanding of these key concepts and definitions in order to fully appreciate the role that metallurgy plays in our daily lives.

CHAPTER 3 | BEST PRACTICES FOR REFRACTORY SELECTION IN HIGH-TEMPERATURE APPLICATIONS

Refractories are materials that are designed to withstand high temperatures and severe conditions such as thermal shock, abrasion, and chemical attack. Refractories are essential components of high-temperature industrial processes such as steelmaking, cement production, and glass manufacturing. The selection of the right refractory material for a particular application is crucial for the successful operation of the process.

Factors to Consider When Selecting Refractory Materials

Several factors need to be considered when selecting refractory materials for high-temperature applications. These include:

1. **Operating temperature:** The operating temperature is the most critical factor in selecting the appropriate refractory material. The material must be able to withstand the maximum temperature of the process without degrading or failing.

2. **Thermal shock resistance:** Refractory materials are exposed to rapid temperature changes, which can cause cracking and failure. The refractory material must have high thermal shock resistance to withstand these conditions.

3. **Abrasion resistance:** In some applications, refractory materials are subjected to abrasive wear from the materials being processed. The material must have high abrasion resistance to withstand this wear.

4. **Chemical resistance:** Refractory materials are often exposed to aggressive chemicals in high-temperature applications. The material must have good chemical resistance to prevent corrosion and degradation.

5. **Cost:** The cost of the refractory material is an important consideration in selecting the appropriate material. The material must provide a cost-effective solution while still meeting the required specifications.

Types of Refractory Materials

There are several types of refractory materials available for high-temperature applications, including:

Fireclay refractories: These are made from clay and are suitable for low-temperature applications up to 1,550°C.

Silica refractories: These are made from pure silica and are suitable for high-temperature applications up to 1,800°C.

High-alumina refractories: These are made from alumina and are suitable for high-temperature applications up to 1,800°C.

Magnesia refractories: These are made from magnesia and are suitable for high-temperature applications up to 2,000°C.

Carbon-based refractories: These are made from carbon and are suitable for high-temperature applications up to 2,500°C.

Best Practices for Refractory Selection

To ensure the selection of the appropriate refractory material, the following best practices should be considered:

1. Conduct a thorough analysis of the process requirements and the operating conditions to determine the necessary properties of the refractory material.

2. Consult with refractory experts and suppliers to determine the most suitable material for the application.

3. Consider the overall cost of the refractory material, including installation and maintenance costs, when selecting the material.

4. Use the correct installation techniques to ensure the proper installation of the refractory material and prevent premature failure.

5. Conduct regular inspections and maintenance to ensure the continued performance of the refractory material.

Conclusion

The selection of the appropriate refractory material is crucial for the successful operation of high-temperature industrial processes. By considering the factors discussed in this chapter and following the best practices for refractory selection, industrial operations can ensure the long-term performance and efficiency of their refractory materials.

CHAPTER 4 | Case Studies in Refractory Failures: Lessons Learned

Refractory materials play a critical role in high-temperature industrial applications. They are used to line furnaces, kilns, reactors, and other equipment that operates at extreme temperatures. Refractory materials must be able to withstand the thermal shock and corrosive environments of these applications while maintaining their mechanical strength and structural integrity. However, despite their importance, refractory failures are not uncommon. In this chapter, we will explore several case studies of refractory failures and the lessons learned from them.

Case Study 1: Refractory Failure in a Cement Kiln

In this case study, a cement kiln was experiencing refractory failure in the burning zone. The refractory lining was cracking and spalling, leading to reduced production and increased maintenance costs. Investigation revealed that the failure was due to alkali attack on the refractory material. The high-alkali content of the raw materials was causing a reaction with the refractory material, leading to its degradation. The lesson learned from this case is the importance of understanding the chemical composition of the raw materials and their potential impact on the refractory material.

Case Study 2: Refractory Failure in a Steel Ladle

In this case study, a steel ladle was experiencing refractory failure, resulting in slag buildup and reduced heat retention. Investigation revealed that the failure was due to thermal shock. The temperature differential between the hot steel and the cold ladle was causing rapid expansion and contraction of the refractory material, leading to its cracking and failure. The lesson learned from this case is the importance of proper thermal design and insulation of the equipment to minimize thermal shock.

Case Study 3: Refractory Failure in a Glass Furnace

In this case study, a glass furnace was experiencing refractory failure in the crown and sidewalls. Investigation revealed that the failure was due to glass infiltration into the refractory material. The high-temperature molten glass was penetrating the refractory material, leading to its erosion and degradation. The lesson learned from this case is the importance of proper selection and installation of the refractory material to resist chemical attack and erosion.

Case Study 4: Refractory Failure in a Petrochemical Reactor

In this case study, a petrochemical reactor was experiencing refractory failure in the catalyst bed. The refractory lining was eroding, leading to reduced catalyst activity and increased reactor pressure drop. Investigation revealed that the failure was due to the presence of corrosive gases in the reactor. The refractory material was not resistant to the corrosive environment, leading to its degradation. The lesson learned from this case is the importance of selecting a refractory material that is resistant to the corrosive environment and monitoring its condition regularly.

Conclusion

In conclusion, refractory failures can have significant economic and safety implications in industrial applications. Proper selection, installation, and maintenance of refractory materials are essential to prevent failures and ensure safe and efficient operation of equipment. The case studies discussed in this chapter highlight the importance of understanding the operating conditions and potential failure modes of the refractory material and implementing appropriate measures to mitigate the risks.

Chapter 5 | Understanding the Metallurgical Properties of Steel

Steel is one of the most widely used materials in the world, and for good reason. It is strong, durable, and versatile, making it an ideal choice for a wide range of applications, from construction to manufacturing. But what exactly is steel, and how does it acquire its unique properties? In this chapter, we will explore the metallurgical properties of steel, including its composition, microstructure, and mechanical properties.

Composition of Steel

Steel is an alloy composed primarily of iron and carbon, with other elements added in varying amounts to achieve specific properties. The most common alloying elements added to steel are manganese, nickel, chromium, and molybdenum. These elements can help improve the strength, toughness, and corrosion resistance of the steel.

The amount of carbon in steel also plays a significant role in its properties. Low-carbon steel contains less than 0.3% carbon and is easy to shape and weld. Medium-carbon steel contains between 0.3% and 0.6% carbon and is often used for structural components. High-carbon steel contains more than 0.6% carbon and is very strong, but also more brittle and difficult to shape.

Microstructure of Steel

The microstructure of steel refers to the arrangement of the steel's atoms and molecules, and it can have a significant impact on the material's mechanical properties. Steel can have several different microstructures, including ferrite, pearlite, martensite, and bainite.

Ferrite is a soft, ductile microstructure that is easy to shape but relatively weak. Pearlite is a harder and stronger microstructure that forms when the steel is cooled slowly. Martensite is a very hard and strong microstructure that forms when the steel is cooled rapidly. Bainite is a microstructure that is harder than ferrite but not as hard as martensite.

Mechanical Properties of Steel

The mechanical properties of steel refer to how it behaves when subjected to stress, strain, or deformation. The most important mechanical properties of steel are its strength, ductility, toughness, and hardness.

Strength refers to how much stress a steel material can withstand before it fractures or fails. Ductility refers to how easily a steel material can be stretched or deformed without breaking. Toughness refers to how much energy a steel material can absorb before it fractures or breaks. Hardness refers to how resistant a steel material is to indentation or abrasion.

Conclusion

Understanding the metallurgical properties of steel is essential for anyone working with this material. By understanding the composition, microstructure, and mechanical properties of steel, engineers and designers can select the appropriate steel material for a specific application and ensure that it will perform as expected.

CHAPTER 6 | Innovations in Steelmaking Technology: Current and Future Developments

Steel has been the backbone of the world's industrial and economic development for centuries. Over the years, there have been numerous advancements in steelmaking technology that have made it possible to produce stronger, more durable, and higher quality steel products. In this chapter, we will explore the latest innovations in steelmaking technology and discuss how they are changing the industry.

Electric Arc Furnace (EAF)

The electric arc furnace (EAF) is a type of furnace used for steelmaking that uses an electric arc to melt down scrap steel. The EAF is a highly efficient and flexible steelmaking process that allows for rapid changes in steel production and has a lower environmental impact than traditional blast furnaces. Advancements in EAF technology have made it possible to produce high-quality steel with a lower carbon footprint.

Laser-based Steelmaking

Laser-based steelmaking is a new technology that uses lasers to melt and shape steel. This process allows for greater precision and control over the steel's composition and properties, resulting in higher quality steel products. The use of lasers also reduces the amount of energy required to produce steel, making it a more sustainable option.

Oxygen Steelmaking

Oxygen steelmaking is a steelmaking process that uses pure oxygen to increase the temperature and speed of the reaction between the iron and other alloying elements. This process reduces the amount of time required to produce steel and results in higher-quality steel products. Advancements in oxygen steelmaking technology have made it possible to produce steel with a much lower carbon footprint.

Hydrogen Steelmaking

Hydrogen steelmaking is a new steelmaking process that uses hydrogen instead of carbon to reduce the iron ore. This process produces water instead of carbon dioxide, resulting in a much lower carbon footprint. Advancements in hydrogen steelmaking technology are still ongoing, but it has the potential to revolutionize the steelmaking industry and make it even more sustainable.

Conclusion

Innovations in steelmaking technology have transformed the industry over the years and continue to do so. From electric arc furnaces to laser-based steelmaking, the latest advancements are focused on producing higher-quality steel with a lower carbon footprint. As the industry moves towards a more sustainable future, the adoption of these new technologies will be critical in reducing emissions and ensuring a more environmentally friendly steel production process.

Chapter 7 | Improving the Quality of Steel: Best Practices for Caster Operations

Caster operations are a critical part of the steelmaking process, where liquid steel is solidified into solid steel products of different shapes and sizes. The quality of the steel produced in caster operations depends on several factors, including the equipment used, process control, and operator skills. In this chapter, we will explore best practices for improving the quality of steel in caster operations.

Equipment Selection

The selection of the right equipment is critical to the quality of steel produced in caster operations. The selection of the caster type, mold design, and casting speed are crucial parameters that must be optimized to produce high-quality steel. The use of advanced technologies such as electromagnetic stirring, mold oscillation, and mold level control can significantly improve the quality of steel.

Process Control

Process control is essential in caster operations to ensure that the steel produced meets the required quality standards. The use of real-time monitoring systems, sensors, and feedback controls can help operators optimize the casting process and reduce the variability in the quality of the steel produced. Online process control can help reduce downtime, improve productivity, and increase the quality of steel produced.

Operator Training

Operator training is another critical factor in improving the quality of steel produced in caster operations. Operators must have a deep understanding of the casting process and be able to identify potential issues that may affect the quality of the steel produced. Training should cover topics such as mold lubrication, mold powder application, mold inspection, and mold maintenance.

Continuous Improvement

Continuous improvement is key to improving the quality of steel in caster operations. Regular monitoring of the process parameters, analyzing data, and implementing corrective actions can help identify and eliminate issues that affect the quality of steel produced. Continuous improvement can help reduce defects, increase yield, and improve the overall quality of the steel produced.

Conclusion

Improving the quality of steel in caster operations requires a holistic approach that considers equipment selection, process control, operator training, and continuous improvement. The adoption of advanced technologies and the use of real-time monitoring systems can significantly improve the quality of steel produced. Training operators and encouraging continuous improvement can help reduce defects, increase yield, and improve the overall quality of the steel produced. By following best practices, caster operations can produce high-quality steel products that meet the stringent quality standards of the industry.

Chapter 8 | The Role of Metallurgy in the Aerospace Industry

The aerospace industry is one of the most demanding industries in terms of material performance, reliability, and safety. The selection of materials for aerospace applications is critical, and metallurgy plays a vital role in meeting these requirements. In this chapter, we will explore the role of metallurgy in the aerospace industry.

Material Selection

The selection of materials for aerospace applications is driven by factors such as strength, weight, corrosion resistance, and temperature resistance. The use of advanced alloys, composites, and coatings can significantly improve the performance of aerospace components. The selection of the right material requires a deep understanding of the properties of the material, the operating conditions, and the manufacturing process.

Manufacturing Processes

The manufacturing process of aerospace components must meet the highest standards of quality and reliability. Metallurgical processes such as forging, casting, and welding are critical to the manufacturing of aerospace components. The use of advanced manufacturing techniques such as additive manufacturing can significantly reduce lead times and improve the quality of the components.

Testing and Certification

Testing and certification are critical steps in ensuring that aerospace components meet the required performance, reliability, and safety standards. Metallurgical testing techniques such as microstructure analysis, tensile testing, and fatigue testing are used to evaluate the properties of materials and components. The use of non-destructive testing techniques such as X-ray, ultrasound, and eddy current testing can detect defects that may affect the performance of components.

Future Developments

The aerospace industry is continuously evolving, and new materials and manufacturing processes are being developed to meet the increasing demands for performance and safety. The use of advanced materials such as nanomaterials and smart materials can significantly improve the performance of aerospace components. The use of artificial intelligence and machine learning can help optimize the manufacturing process and reduce defects.

Conclusion

Metallurgy plays a vital role in the aerospace industry, from the selection of materials to the manufacturing of components and testing and certification. The aerospace industry demands the highest standards of performance, reliability, and safety, and metallurgy is essential in meeting these requirements. The adoption of advanced materials, manufacturing processes, and testing techniques can significantly improve the performance of aerospace components. The future of the aerospace industry will continue to rely on metallurgy to meet the increasingly demanding requirements for performance and safety.

Chapter 9 | Corrosion Prevention and Control in Metallurgical Applications

Corrosion is a natural process that can have significant economic and safety impacts in metallurgical applications. Corrosion prevention and control are critical in maintaining the integrity and longevity of metallurgical components. In this chapter, we will explore the various methods for preventing and controlling corrosion in metallurgical applications.

Corrosion Mechanisms

Corrosion is an electrochemical process that involves the transfer of electrons from one material to another. The most common types of corrosion in metallurgical applications are uniform corrosion, pitting corrosion, crevice corrosion, and stress corrosion cracking. The selection of the appropriate corrosion prevention and control method depends on the type of corrosion mechanism and the environment in which the component is used.

Corrosion Prevention and Control Methods

Several methods can be used for preventing and controlling corrosion in metallurgical applications. These methods include:

1. Protective Coatings: Coatings such as paints, lacquers, and enamels can protect the metal surface from the environment.

2. Cathodic Protection: Cathodic protection involves applying a sacrificial anode to the metal surface, which corrodes instead of the metal.

3. Corrosion Inhibitors: Corrosion inhibitors are chemicals that can be added to the environment or applied to the metal surface to prevent corrosion.

4. Material Selection: The selection of materials with a high resistance to corrosion can significantly reduce the risk of corrosion.

5. Environmental Control: Controlling the environment in which the component is used can prevent or reduce the rate of corrosion.

6. Maintenance and Inspection: Regular maintenance and inspection can detect corrosion early and prevent further damage.

Conclusion

Corrosion prevention and control are critical in metallurgical applications, as corrosion can have significant economic and safety impacts. The selection of the appropriate corrosion prevention and control method depends on the type of corrosion mechanism and the environment in which the component is used. Protective coatings, cathodic protection, corrosion inhibitors, material selection, environmental control, and regular maintenance and inspection are effective methods for preventing and controlling corrosion. Metallurgists and engineers must understand the principles of corrosion prevention and control to ensure the integrity and longevity of metallurgical components.

Chapter 10 | The Role of Metallurgy in the Automotive Industry

Metallurgy plays a crucial role in the automotive industry, as the industry relies heavily on high-performance metals and alloys to ensure the safety and performance of vehicles. In this chapter, we will explore the importance of metallurgy in the automotive industry and the different metals and alloys used in automotive applications.

Importance of Metallurgy in the Automotive Industry

Metallurgy is critical to the automotive industry as it provides the materials needed for the construction of engine components, chassis, and other critical automotive parts. The use of high-performance metals and alloys is essential to ensure the safety and performance of vehicles. Metallurgists work closely with automotive engineers to develop new alloys and processes that can improve the strength, durability, and performance of automotive components.

Metals and Alloys Used in Automotive Applications

1. Steel: Steel is one of the most commonly used metals in the automotive industry, as it is strong, durable, and affordable. High-strength steels are used in critical parts such as the frame, suspension, and body panels.

2. Aluminum: Aluminum is lightweight and corrosion-resistant, making it an ideal material for automotive applications. It is commonly used in engine components, wheels, and body panels.

3. Magnesium: Magnesium is lightweight and has excellent mechanical properties, making it an ideal material for automotive applications. It is commonly used in engine components, transmission housings, and steering wheel frames.

4. Titanium: Titanium is a high-strength and lightweight metal that is used in high-performance automotive applications such as race cars and high-end sports cars.

5. Copper: Copper is used in automotive applications for electrical wiring and as a component in radiators and heat exchangers.

6. Nickel: Nickel is used in the production of stainless steel, which is commonly used in automotive exhaust systems and other high-temperature applications.

Innovations in Metallurgy for the Automotive Industry

Metallurgists are continuously developing new alloys and processes to improve the performance of automotive components. Recent innovations include:

1. Advanced High-Strength Steels: These steels offer increased strength and reduced weight compared to conventional high-strength steels.

2. Lightweight Aluminum Alloys: New aluminum alloys are being developed that are stronger and more durable than existing alloys, allowing for even greater weight reduction in automotive components.

3. Magnesium Alloys: Advances in magnesium alloy technology are increasing the use of this lightweight metal in automotive applications.

4. Advanced Coatings: New coatings are being developed that can protect automotive components from corrosion and wear.

Conclusion

Metallurgy plays a vital role in the automotive industry, providing the materials needed for the construction of engine components, chassis, and other critical automotive parts. The use of high-performance metals and alloys is essential to ensure the safety and performance of vehicles. Metallurgists work closely with automotive engineers to develop new alloys and processes that can improve the strength, durability, and performance of automotive components. Advances in metallurgy are continuously improving the performance and efficiency of vehicles, leading to safer and more sustainable transportation.

Chapter 11 | The Art and Science of Heat Treatment: Key Principles and Techniques

Heat treatment is a critical process in metallurgy that involves heating and cooling metal to alter its mechanical properties. The art and science of heat treatment is a combination of metallurgical principles and practical techniques. In this chapter, we will explore the key principles and techniques of heat treatment.

Key Principles of Heat Treatment

1. Heating: The first step in heat treatment is heating the metal to a specific temperature. The temperature depends on the type of metal and the desired mechanical properties.

2. Holding: Once the metal reaches the desired temperature, it is held at that temperature for a specific amount of time. This is known as the soaking period.

3. Cooling: After the holding period, the metal is cooled in a controlled manner to achieve the desired mechanical properties.

4. Annealing: Annealing is a heat treatment process that involves heating the metal to a specific temperature and holding it there for an extended period. This process helps to relieve internal stresses and improve the ductility of the metal.

5. Tempering: Tempering is a heat treatment process that involves heating the metal to a specific temperature and then cooling it in a controlled manner. This process helps to reduce the hardness of the metal and increase its toughness.

Techniques of Heat Treatment

1. Normalizing: Normalizing is a heat treatment process that involves heating the metal to a specific temperature and holding it there for a specific amount of time. The metal is then cooled in air. This process helps to refine the grain structure of the metal and improve its toughness.

2. Hardening: Hardening is a heat treatment process that involves heating the metal to a specific temperature and then cooling it in a controlled manner to achieve the desired mechanical properties. The metal is then quenched in oil or water to rapidly cool it.

3. Case hardening: Case hardening is a heat treatment process that involves heating the metal to a specific temperature and then diffusing carbon into the surface of the metal. This process creates a hard outer layer while maintaining a tough, ductile core.

4. Nitriding: Nitriding is a heat treatment process that involves diffusing nitrogen into the surface of the metal. This process creates a hard outer layer while maintaining a tough, ductile core.

5. Martempering: Martempering is a heat treatment process that involves quenching the metal in a warm bath. This process reduces the risk of distortion and cracking, while still achieving the desired mechanical properties.

Conclusion

Heat treatment is a critical process in metallurgy that involves heating and cooling metal to alter its mechanical properties. The art and science of heat treatment is a combination of metallurgical principles and practical techniques. The key principles of heat treatment include heating, holding, and cooling the metal, as well as annealing and tempering. The techniques of heat treatment include normalizing, hardening, case hardening, nitriding, and martempering. Understanding the principles and techniques of heat treatment is essential for achieving the desired mechanical properties in metal components.

Chapter 12 | The Role of Refractory Materials in Glass Manufacturing

Refractory materials play a crucial role in glass manufacturing, as they provide the necessary resistance to high temperatures, chemical corrosion, and mechanical wear. In this chapter, we will explore the key characteristics of refractory materials used in glass manufacturing and their role in the glass-making process.

Key Characteristics of Refractory Materials

1. High temperature resistance: Refractory materials must be able to withstand high temperatures, as glass is melted at temperatures ranging from 1400°C to 1600°C.

2. Chemical resistance: Refractory materials must resist chemical corrosion from the glass melt and any additives used in the glass-making process.

3. Mechanical strength: Refractory materials must be able to withstand mechanical wear caused by the flow of the glass melt and any mechanical equipment used in the glass-making process.

4. Thermal shock resistance: Refractory materials must be able to resist thermal shock caused by the rapid heating and cooling of the glass melt.

Role of Refractory Materials in Glass Manufacturing

1. Furnace lining: Refractory materials are used to line the melting furnaces used in glass manufacturing. The lining must be able to withstand high temperatures and chemical corrosion from the glass melt.

2. Feeder channels: Refractory materials are used to create the feeder channels that allow the glass melt to flow from the melting furnace to the forming process. These channels must be able to withstand high temperatures and resist mechanical wear.

3. Glass contact areas: Refractory materials are used to create the molds, plungers, and other equipment used to form the glass products. These materials must be able to resist chemical corrosion and mechanical wear from the glass melt.

4. Regenerator lining: Refractory materials are used to line the regenerators, which are used to preheat the combustion air before it enters the melting furnace. These materials must be able to withstand high temperatures and resist thermal shock.

Conclusion

Refractory materials are an essential component of glass manufacturing, as they provide the necessary resistance to high temperatures, chemical corrosion, and mechanical wear. The key characteristics of refractory materials used in glass manufacturing include high temperature resistance, chemical resistance, mechanical strength, and thermal shock resistance. Refractory materials are used in furnace linings, feeder channels, glass contact areas, and regenerator linings. Understanding the role of refractory materials in glass manufacturing is essential for producing high-quality glass products.

CHAPTER 13 | METALLURGICAL ANALYSIS TECHNIQUES: A GUIDE FOR BEGINNERS

Metallurgical analysis techniques are used to determine the physical, chemical, and mechanical properties of metals and alloys. In this chapter, we will explore some of the most commonly used metallurgical analysis techniques, including their principles, applications, advantages, and limitations.

1. Optical Microscopy

Optical microscopy is a non-destructive technique that involves using visible light to examine the microstructure of metals and alloys. It can be used to identify the phases, grain size, and defects present in a material. It is widely used in quality control and failure analysis of metals and alloys.

Advantages: Non-destructive, easy to use, provides visual images.

Limitations: Limited to surface features, may not be able to identify some phases or defects.

2. Scanning Electron Microscopy (SEM)

Scanning electron microscopy is a high-resolution imaging technique that uses an electron beam to scan the surface of a sample. It can provide detailed images of the microstructure and morphology of metals and alloys. It is widely used in materials science research and failure analysis.

Advantages: High-resolution imaging, can provide information on surface features, morphology, and chemical composition.

Limitations: Requires a vacuum environment, may damage the sample, complex instrumentation.

3. X-Ray Diffraction (XRD)

X-ray diffraction is a technique that uses X-rays to determine the crystal structure of metals and alloys. It can be used to identify the phases present in a material and their crystal structures. It is widely used in materials science research and quality control.

Advantages: Can identify crystal structures, non-destructive, easy to use.

Limitations: Limited to crystalline materials, requires a specialized instrument.

4. Differential Scanning Calorimetry (DSC)

Differential scanning calorimetry is a thermal analysis technique that measures the heat flow of a material as it is heated or cooled. It can be used to determine the phase transitions, melting points, and thermal stability of metals and alloys. It is widely used in materials science research and quality control.

Advantages: Can determine phase transitions and melting points, non-destructive, easy to use.

Limitations: Requires a specialized instrument, limited to thermal properties.

5. Transmission Electron Microscopy (TEM)

Transmission electron microscopy is an imaging technique that uses a focused electron beam to pass through a thin section of a sample. It can provide high-resolution images of the microstructure and crystal defects present in metals and alloys. It is widely used in materials science research and failure analysis.

Advantages: High-resolution imaging, can provide information on microstructure, crystal defects, and chemical composition.

Limitations: Requires a vacuum environment and thin sections, complex instrumentation.

Conclusion

Metallurgical analysis techniques are essential for understanding the physical, chemical, and mechanical properties of metals and alloys. Optical microscopy, scanning electron microscopy, X-ray diffraction, differential scanning calorimetry, and transmission electron microscopy are some of the commonly used techniques. Each technique has its advantages and limitations, and the choice of technique depends on the type of material and the information required. Understanding the principles, applications, advantages, and limitations of these techniques is essential for beginners in metallurgical analysis.

Chapter 14 | The Advantages and Disadvantages of Different Steelmaking Processes

Steel is one of the most widely used materials in the world due to its strength, durability, and versatility. Steel can be produced using a variety of different processes, each with its own advantages and disadvantages. In this chapter, we will explore the most common steelmaking processes and their advantages and disadvantages.

1. Basic Oxygen Furnace (BOF)

The basic oxygen furnace is the most common steelmaking process, accounting for over 60% of global steel production. In this process, pig iron is melted in a furnace and then oxygen is blown through the molten metal to remove impurities and adjust the carbon content.

Advantages: High production rates, low energy consumption, good control of carbon content.

Disadvantages: High capital costs, produces large amounts of slag, limited flexibility in product range.

2. Electric Arc Furnace (EAF)

The electric arc furnace is a process that uses an electric arc to melt scrap steel and other raw materials. This process is often used for specialty steels and alloys.

Advantages: Low capital costs, flexibility in raw materials, good control of alloy content.

Disadvantages: High energy consumption, limited production rates, requires large amounts of scrap steel.

3. Induction Furnace (IF)

The induction furnace is a process that uses an electromagnetic field to heat and melt the metal. This process is often used for small-scale production of specialty steels and alloys.

Advantages: Low energy consumption, good control of alloy content, flexibility in raw materials.

Disadvantages: Limited production rates, high capital costs, requires specialized equipment.

4. Vacuum Induction Melting (VIM)

The vacuum induction melting process is a high-precision process used to produce high-quality steels and alloys. In this process, the metal is melted in a vacuum environment to prevent contamination.

Advantages: High quality of the final product, good control of alloy content, low levels of impurities.

Disadvantages: High capital costs, limited production rates, requires specialized equipment.

5. Direct Reduced Iron (DRI)

The direct reduced iron process is a process that converts iron ore into a metallic sponge without melting the metal. The sponge can then be used to produce steel using an electric arc furnace or a basic oxygen furnace.

Advantages: Low energy consumption, flexibility in raw materials, low levels of impurities.

Disadvantages: Limited production rates, requires specialized equipment, high capital costs.

Conclusion

Each steelmaking process has its own advantages and disadvantages, and the choice of process depends on various factors such as the desired product quality, production rates, energy consumption, and capital costs. The basic oxygen furnace and electric arc furnace are the most widely used processes, while induction furnace, vacuum induction melting, and direct reduced iron are used for specialty steels and alloys. Understanding the advantages and disadvantages of these processes is essential for steelmakers to make informed decisions about their production processes.

Chapter 15 | Refractory Design and Installation Best Practices

Refractory materials are essential for many high-temperature industrial processes such as steelmaking, glass manufacturing, and ceramics production. Proper design and installation of refractory materials are critical to ensure the longevity and efficiency of these processes. In this chapter, we will explore the best practices for refractory design and installation.

1. Material Selection

The first step in designing a refractory lining is selecting the appropriate refractory material. The refractory material should be chosen based on the specific operating conditions of the process such as temperature, chemical composition, and abrasiveness of the materials being processed. Common refractory materials include fire clay, high alumina, silica, magnesite, and zircon.

2. Lining Thickness

The thickness of the refractory lining should be sufficient to provide insulation and protection against the operating conditions. The thickness of the lining should be designed based on the temperature, mechanical stresses, and chemical composition of the materials being processed.

3. Anchor Systems

Anchors are used to secure the refractory lining to the metal shell of the process vessel. The anchor system should be designed to accommodate the thermal expansion and contraction of the refractory lining. The selection of the anchor material and spacing should be based on the operating conditions of the process.

4. Installation Techniques

The installation of refractory materials requires specialized knowledge and skills. The installation techniques should be selected based on the type of refractory material being used and the operating conditions of the process. The installation should be performed by trained and experienced personnel to ensure the quality of the installation.

5. Curing

After the refractory lining is installed, it should be cured to ensure the proper bond between the refractory material and the metal shell. The curing time and temperature should be based on the type of refractory material being used and the operating conditions of the process.

6. Maintenance

Proper maintenance of the refractory lining is critical to ensure its longevity and efficiency. Regular inspections should be performed to identify any cracks, erosion, or other damage. Any damage should be repaired immediately to prevent further damage and reduce downtime.

Conclusion

Refractory design and installation best practices are critical to the success of high-temperature industrial processes. The proper selection of refractory materials, lining thickness, anchor systems, installation techniques, curing, and maintenance can ensure the longevity and efficiency of these processes. Refractory lining design and installation require specialized knowledge and skills, and should be performed by trained and experienced personnel.

Chapter 16 | The Role of Metallurgy in Renewable Energy Technologies

The transition towards a sustainable energy system is driven by the increasing demand for clean and renewable energy. The production and deployment of renewable energy technologies require materials that can withstand harsh operating conditions and deliver high performance. Metallurgy plays a crucial role in providing the materials needed for these technologies. In this chapter, we will explore the role of metallurgy in renewable energy technologies.

1. Wind Energy

Wind turbines are a vital component of the renewable energy mix. The manufacture of wind turbine components requires materials that can withstand harsh environmental conditions and high mechanical stresses. Metallurgy provides the materials needed for the production of wind turbine blades, gearboxes, and towers. Advanced materials such as carbon fiber reinforced polymers (CFRP), aluminum alloys, and high-strength steels are used to enhance the performance of wind turbine components.

2. Solar Energy

Solar panels are another key component of the renewable energy mix. The production of solar panels requires materials that can withstand high temperatures and corrosion. Metallurgy provides the materials needed for the production of solar panel frames and supports, as well as the wiring and connectors needed for the electrical connections. Advanced materials such as titanium, stainless steel, and aluminum alloys are commonly used in solar panel manufacturing.

3. Hydrogen Energy

Hydrogen is a clean and versatile fuel that can be produced from renewable energy sources. The production, storage, and transportation of hydrogen require materials that can withstand high temperatures, high pressures, and aggressive chemical environments. Metallurgy provides the materials needed for the construction of hydrogen fuel cells, storage tanks, and pipelines. Advanced materials such as nickel-based alloys, titanium, and stainless steel are used in the production of hydrogen energy technologies.

4. Geothermal Energy

Geothermal energy is a reliable and sustainable source of energy that utilizes the heat from the Earth's interior. The production and deployment of geothermal energy technologies require materials that can withstand high temperatures, corrosion, and abrasion. Metallurgy provides the materials needed for the production of geothermal power plants, including heat exchangers, pipes, and turbines. Advanced materials such as superalloys, nickel-based alloys, and titanium are commonly used in geothermal energy technologies.

Conclusion

The transition towards a sustainable energy system requires the development and deployment of renewable energy technologies. Metallurgy plays a critical role in providing the materials needed for these technologies. Wind energy, solar energy, hydrogen energy, and geothermal energy are just a few examples of the renewable energy technologies that rely on advanced metallurgical materials. The production of these materials requires specialized knowledge and expertise, and the metallurgical industry is continually evolving to meet the demands of the renewable energy sector.

CHAPTER 17 | IMPROVING STEEL PRODUCTION EFFICIENCY: BEST PRACTICES AND STRATEGIES

In an ever-evolving and competitive steel industry, production efficiency is critical to the success and sustainability of steel manufacturers. In this chapter, we will explore best practices and strategies for improving steel production efficiency.

1. Optimize Raw Material Selection

Steel manufacturers can improve production efficiency by optimizing the selection of raw materials. The quality of raw materials, such as iron ore and coal, can impact the efficiency of the steel production process. Steel manufacturers can improve efficiency by selecting raw materials with consistent quality and composition.

2. Implement Lean Manufacturing Practices

Lean manufacturing is a systematic approach to identifying and eliminating waste in the production process. Steel manufacturers can implement lean manufacturing practices to reduce production costs and improve efficiency. Lean manufacturing practices include identifying and eliminating non-value-added activities, optimizing production flows, and implementing just-in-time inventory management.

3. Utilize Advanced Process Control Technologies

Advanced process control technologies can improve the efficiency of steel production processes. These technologies include real-time process monitoring, process modeling and simulation, and optimization algorithms. By utilizing advanced process control technologies, steel manufacturers can reduce production costs and improve process efficiency.

4. Implement Predictive Maintenance Strategies

Steel manufacturers can improve production efficiency by implementing predictive maintenance strategies. Predictive maintenance involves monitoring equipment and systems to identify potential issues before they cause downtime or production delays. By implementing predictive maintenance strategies, steel manufacturers can reduce equipment downtime, extend the lifespan of equipment, and improve production efficiency.

5. Invest in Employee Training and Development

Investing in employee training and development is essential for improving steel production efficiency. By providing employees with training and development opportunities, steel manufacturers can improve employee skillsets and increase employee engagement. This can lead to a more efficient and productive workforce.

Conclusion

Improving steel production efficiency is critical for the success and sustainability of steel manufacturers. By optimizing raw material selection, implementing lean manufacturing practices, utilizing advanced process control technologies, implementing predictive maintenance strategies, and investing in employee training and development, steel manufacturers can improve production efficiency, reduce costs, and stay competitive in a rapidly evolving industry.

Chapter 18 | The Importance of Alloy Design in Metallurgy

Alloy design plays a crucial role in metallurgy. Metallurgists design alloys to achieve desired properties, such as strength, corrosion resistance, and ductility, for specific applications. In this chapter, we will explore the importance of alloy design in metallurgy.

1. Tailoring Properties to Application Requirements

The primary purpose of alloy design is to tailor the properties of a material to meet the requirements of specific applications. For example, the design of high-strength alloys is crucial in the aerospace and automotive industries, where lightweight materials with high strength and durability are essential. Similarly, the design of corrosion-resistant alloys is crucial in chemical processing plants and marine environments.

2. Balancing Strength and Ductility

In many applications, it is essential to balance the strength and ductility of an alloy. An alloy that is too brittle may fracture under load, while an alloy that is too ductile may deform and fail prematurely. Metallurgists use alloy design to balance the strength and ductility of an alloy to achieve the desired mechanical properties.

3. Optimizing Manufacturing Processes

Alloy design also plays a crucial role in optimizing manufacturing processes. By designing alloys with specific properties, metallurgists can develop processes that are more efficient, cost-effective, and environmentally friendly. For example, the design of high-temperature alloys has led to the development of more efficient gas turbine engines, which reduce fuel consumption and emissions.

4. Exploring New Applications

Alloy design is essential in exploring new applications for materials. By designing alloys with unique properties, metallurgists can explore new areas of application and develop new technologies. For example, the design of shape memory alloys has led to the development of new medical devices, such as stents and orthodontic wires.

5. Advancing the Field of Metallurgy

Alloy design is at the forefront of advancing the field of metallurgy. By continuously developing new alloys and refining existing ones, metallurgists can push the boundaries of what is possible in material science. This has led to numerous technological advancements in a wide range of industries, including aerospace, automotive, and medical.

Conclusion

Alloy design is a crucial aspect of metallurgy that enables the development of materials with specific properties for specific applications. By tailoring the properties of materials to meet application requirements, balancing strength and ductility, optimizing manufacturing processes, exploring new applications, and advancing the field of metallurgy, metallurgists can push the boundaries of what is possible in material science and contribute to the development of new technologies in various industries.

Chapter 19 | The Challenges and Opportunities of Metallurgy in the 21st Century

Metallurgy has been an essential field of study and practice for centuries. It has played a crucial role in the development of human civilization and has enabled the creation of some of the world's most significant structures and machines. However, with the emergence of new technologies and changing global trends, the field of metallurgy is facing both challenges and opportunities in the 21st century. This chapter will discuss some of the significant challenges and opportunities facing the field of metallurgy in the current era.

1. Sustainable Metallurgy:

The demand for metals and materials continues to rise globally, and with it, the environmental impact of metallurgical processes. Sustainable metallurgy is the need of the hour, and metallurgists need to develop new techniques and technologies that are more energy-efficient, reduce carbon emissions, and minimize waste.

2. Advanced Materials:

Metallurgists are now using advanced materials to develop new products that are stronger, lighter, and more durable. These materials are revolutionizing industries such as aerospace, defense, and automotive. However, creating these advanced materials requires a deep understanding of the properties of various metals and alloys.

3. Digitization:

The field of metallurgy is being transformed by digital technologies such as machine learning, artificial intelligence, and big data. These technologies are enabling the creation of predictive models that can simulate metallurgical processes and predict the properties of materials before they are manufactured. This will help to reduce the time and costs associated with developing new materials and processes.

4. Globalization:

The globalization of the metallurgical industry is bringing new opportunities for growth and collaboration. Metallurgists are now working across borders to develop new technologies, share best practices, and expand their knowledge. However, the globalization of the industry also poses challenges such as different regulations, standards, and cultural differences.

5. Workforce Development:

The metallurgical industry is facing a shortage of skilled workers, and there is a need to attract and retain a new generation of metallurgists. The field of metallurgy is also evolving, and there is a need to provide continuous training and development opportunities to keep up with the changing demands of the industry.

Conclusion:

The field of metallurgy is facing significant challenges and opportunities in the 21st century. Sustainable metallurgy, advanced materials, digitization, globalization, and workforce development are some of the key areas where the industry needs to focus. Metallurgists need to adapt to these changes, embrace new technologies, and work collaboratively to drive innovation and growth in the field.

Chapter 20 | Understanding the Role of Refractory Materials in Cement Manufacturing

Cement manufacturing is a complex process that involves high-temperature reactions and the use of various chemicals. The process of cement manufacturing involves the heating of raw materials to a high temperature, which results in the formation of clinker. The clinker is then ground to a fine powder and mixed with other materials to form cement. Refractory materials play a crucial role in the cement manufacturing process, as they are used to line the high-temperature equipment and prevent damage to the equipment.

Refractory Materials Used in Cement Manufacturing:

Various types of refractory materials are used in cement manufacturing. These materials include high alumina, fireclay, silica, magnesia, and chrome refractories. High alumina and fireclay refractories are commonly used in the cement industry due to their high resistance to abrasion and thermal shock. Silica refractories are used in the upper portion of the cement kiln, where the temperature is relatively low. Magnesia refractories are used in the burning zone of the kiln due to their high resistance to alkalis. Chrome refractories are used in the clinker cooler due to their high resistance to abrasion.

Refractory Installation:

The installation of refractory materials in the cement industry requires careful planning and execution. The selection of the appropriate refractory materials and the correct installation techniques can significantly increase the service life of the refractory lining. The refractory installation process involves several steps, including surface preparation, mixing and placement of refractory materials, and curing.

Refractory Maintenance:

Refractory maintenance is an essential aspect of the cement manufacturing process. Regular inspection and maintenance of the refractory lining can prevent unexpected downtime and minimize repair costs. The maintenance process involves inspection of the refractory lining, repair of any damage or cracks, and replacement of worn-out refractory materials.

Conclusion:

Refractory materials play a crucial role in the cement manufacturing process. The selection, installation, and maintenance of the refractory lining are critical for the efficient operation of the cement plant. Understanding the properties of different refractory materials and their application in the cement industry can significantly improve the service life of the refractory lining and reduce the maintenance cost.

Chapter 21 | Metallurgical Failure Analysis: Common Causes and Prevention Strategies

Metallurgical failure analysis involves identifying the root causes of material failure in metallurgical components or structures. Metallurgical failures can occur due to various reasons, including material defects, manufacturing errors, environmental factors, and service conditions. Failure analysis is an essential tool in identifying and preventing future failures, improving design, and enhancing material performance.

Common Causes of Metallurgical Failures:

Metallurgical failures can result from various factors such as:

1. Material Defects: Material defects such as inclusions, voids, cracks, and discontinuities can lead to premature failure of the component.

2. Manufacturing Errors: Defects arising from manufacturing errors such as improper heat treatment, incorrect welding techniques, or insufficient material processing can lead to metallurgical failures.

3. Environmental Factors: Environmental factors such as temperature, corrosion, wear, and fatigue can cause metallurgical failures.

4. Service Conditions: Service conditions such as loading, impact, vibration, or cyclic loading can result in metallurgical failures.

Prevention Strategies:

Prevention strategies can include:

1. Material Selection: Proper material selection can help prevent material defects, improve performance, and increase component lifespan.

2. Design Optimization: Design optimization can minimize the impact of environmental and service conditions and improve the load-bearing capacity of components.

3. Quality Control: Quality control measures can detect and eliminate manufacturing errors, reduce material defects, and ensure that materials are processed correctly.

4. Maintenance and Inspection: Regular maintenance and inspection can help identify and prevent failures caused by environmental factors or service conditions.

Conclusion:

Metallurgical failure analysis plays a vital role in improving the quality and reliability of metallurgical components and structures. Understanding the common causes of metallurgical failures and implementing prevention strategies can enhance material performance, reduce the risk of failure, and increase the lifespan of components.

Chapter 22 | The Role of Metallurgy in Medical Device Manufacturing

Metallurgical materials are widely used in the medical device industry, where biocompatibility, corrosion resistance, and durability are essential requirements. The metallurgical properties of these materials must meet specific standards and regulations to ensure the safety and efficacy of medical devices. This chapter aims to explore the critical role of metallurgy in medical device manufacturing.

Metallurgical Materials in Medical Device Manufacturing:

Metallurgical materials used in medical device manufacturing can include:

1. Stainless Steel: Stainless steel is commonly used in medical device manufacturing due to its corrosion resistance, high strength, and biocompatibility.

2. Titanium: Titanium is also widely used in the medical industry due to its biocompatibility, corrosion resistance, and low density.

3. Cobalt-Chromium Alloys: Cobalt-chromium alloys are used in orthopedic implants due to their high strength, biocompatibility, and corrosion resistance.

4. Nickel-Titanium Alloys: Nickel-titanium alloys, also known as nitinol, are used in stents, orthopedic implants, and dental applications due to their shape memory and superelasticity.

Metallurgical Processes in Medical Device Manufacturing:

Metallurgical processes used in medical device manufacturing can include:

1. Heat Treatment: Heat treatment processes such as annealing, quenching, and tempering are used to improve the mechanical properties of metallurgical materials used in medical devices.

2. Surface Treatment: Surface treatment processes such as polishing, electroplating, and anodizing are used to improve the corrosion resistance, biocompatibility, and aesthetics of medical devices.

3. Welding: Welding techniques such as laser welding, TIG welding, and resistance welding are used to join metal components used in medical devices.

4. Additive Manufacturing: Additive manufacturing, also known as 3D printing, is increasingly used in the medical device industry to produce complex and customized components.

Regulatory Requirements:

Medical devices must comply with various regulations and standards, including biocompatibility, sterility, and material performance. Regulatory requirements can include ISO 10993 for biocompatibility, ASTM F75 for cobalt-chromium alloys, and ISO 5832-2 for titanium alloys.

Conclusion:

Metallurgy plays a vital role in the medical device industry, where biocompatibility, corrosion resistance, and durability are essential requirements. Understanding the metallurgical properties of materials used in medical device manufacturing and implementing appropriate metallurgical processes can ensure the safety and efficacy of medical devices. Compliance with regulatory requirements is also essential to ensure that medical devices meet the necessary standards for quality and safety.

Chapter 23 | Advanced Steelmaking Techniques: From Microalloying to Powder Metallurgy

In recent years, the steel industry has witnessed significant advancements in steelmaking techniques, which have led to the production of high-quality steel with improved mechanical properties. Two such techniques are microalloying and powder metallurgy, which have revolutionized the steelmaking process. Microalloying involves the addition of small amounts of alloying elements to the steel, while powder metallurgy involves the production of steel from powders. This chapter discusses the principles and advantages of these advanced steelmaking techniques.

Microalloying:

Microalloying is a technique in which small amounts of alloying elements such as niobium, vanadium, and titanium are added to the steel. These elements form carbides and nitrides during the steelmaking process, which improve the strength and toughness of the steel. The chapter discusses the principles and advantages of microalloying, as well as the challenges associated with the technique.

Powder Metallurgy:

Powder metallurgy is a technique in which steel is produced from powders of iron, carbon, and alloying elements. The powders are mixed, compacted, and sintered at high temperatures to form a solid steel. This technique has several advantages, including the production of high-strength and high-toughness steels. The chapter discusses the principles of powder metallurgy, the advantages of the technique, and the challenges associated with the process.

Other Advanced Steelmaking Techniques:

In addition to microalloying and powder metallurgy, there are several other advanced steelmaking techniques that have been developed in recent years. These include high-pressure gas quenching, continuous casting, and hot rolling. The chapter discusses the principles and advantages of these techniques, as well as their limitations and challenges.

Applications of Advanced Steelmaking Techniques:

Advanced steelmaking techniques have found several applications in the industry. The chapter discusses the applications of microalloying, powder metallurgy, and other advanced steelmaking techniques in various industries such as automotive, aerospace, and construction. The chapter also discusses the potential for further developments and innovations in the field of advanced steelmaking techniques.

Conclusion:

Advanced steelmaking techniques such as microalloying and powder metallurgy have revolutionized the steel industry by producing high-quality steel with improved mechanical properties. These techniques have found several applications in various industries and have the potential for further developments and innovations. The chapter highlights the principles, advantages, challenges, and potential of advanced steelmaking techniques.

Chapter 24 | The Importance of Quality Control in Steel Manufacturing

The quality control of steel manufacturing is a crucial aspect that affects the overall success of the industry. Steel is used in various applications, ranging from construction to transportation and energy, and its quality is directly related to the safety and reliability of these products. In this chapter, we will discuss the importance of quality control in steel manufacturing and the various methods used for ensuring high-quality steel production.

Quality Control Methods:

Quality control in steel manufacturing involves a range of methods and techniques to ensure that the finished product meets the required specifications. Some of the most commonly used methods include:

1. Chemical Analysis: Chemical analysis is a crucial method for determining the composition of steel. It involves analyzing the steel sample for the presence of impurities, such as sulphur and phosphorus, that can affect the steel's quality.

2. Mechanical Testing: Mechanical testing is another important method used for quality control in steel manufacturing. It involves testing the steel's mechanical properties, such as tensile strength and toughness, to ensure that they meet the required specifications.

3. Non-Destructive Testing: Non-destructive testing is a method used for detecting defects in steel without causing any damage to the material. Some of the most commonly used non-destructive testing methods include ultrasonic testing, magnetic particle testing, and eddy current testing.

4. Visual Inspection: Visual inspection involves visually examining the steel for any defects, such as cracks or surface imperfections.

5. Quality Management Systems: Quality management systems are a set of procedures and processes used to ensure that the steel production process meets the required quality standards. This includes the establishment of quality objectives, documentation, and regular audits.

Importance of Quality Control:

The importance of quality control in steel manufacturing cannot be overstated. Steel is used in critical applications where failure can have severe consequences. For example, if steel used in the construction of a building is of poor quality, it can result in structural failure and loss of life. Similarly, if steel used in the construction of a bridge is of poor quality, it can result in a catastrophic failure.

Quality control also plays a critical role in ensuring the economic viability of the steel industry. Poor-quality steel can lead to costly repairs, recalls, and reputational damage, resulting in lost revenue and market share.

Conclusion

In conclusion, quality control is a crucial aspect of steel manufacturing that should not be overlooked. The various methods and techniques used for quality control ensure that the finished product meets the required specifications, thereby ensuring its safety and reliability. The importance of quality control extends beyond ensuring product safety; it also plays a critical role in the economic viability of the steel industry. As such, quality control should be an integral part of the steel production process.

Chapter 25 | Exploring the Applications of Powder Metallurgy in Industry

Powder metallurgy is a widely used manufacturing technique that involves the formation of metal powders into useful shapes and components. The process involves a series of steps, including powder production, mixing, compaction, and sintering. Powder metallurgy has many advantages over traditional manufacturing methods, such as the ability to create complex shapes and parts with consistent quality, high purity, and high strength. This chapter will explore the various applications of powder metallurgy in industry.

Powder Production Techniques:

This section will discuss the different methods used for powder production, including mechanical methods such as ball milling, attrition milling, and cryomilling, as well as chemical methods such as chemical reduction, electrolysis, and spray drying.

Powder Mixing and Compaction:

This section will focus on the various techniques used to mix metal powders with other powders, binders, and lubricants to achieve the desired properties. Compaction techniques such as cold pressing, isostatic pressing, and hot pressing will also be discussed.

Sintering and Post-Sintering Processes:

This section will cover the sintering process, which involves heating the compacted powder to a high temperature in a controlled atmosphere to promote bonding between the particles. Post-sintering processes such as heat treatment, hot isostatic pressing, and machining will also be discussed.

Applications of Powder Metallurgy:

This section will explore the various applications of powder metallurgy in different industries, including automotive, aerospace, medical, and electronics. The advantages and limitations of powder metallurgy in each industry will be discussed, along with specific examples of components and products made using powder metallurgy.

Future Developments in Powder Metallurgy:

This section will discuss current and future research in powder metallurgy, including advancements in powder production techniques, novel materials, and emerging applications. The challenges and opportunities for powder metallurgy in the future will also be explored.

Conclusion

Powder metallurgy is a versatile and reliable manufacturing technique that has numerous applications in various industries. This chapter has provided an overview of the different steps involved in the powder metallurgy process, as well as the advantages and limitations of this technique. The examples provided demonstrate the potential of powder metallurgy to create complex, high-quality components that meet the demands of modern industry. Ongoing research in powder metallurgy promises to unlock new possibilities for this innovative manufacturing technique in the future.

CHAPTER 26 | Metallurgical Applications in the Oil and Gas Industry

The oil and gas industry is one of the largest and most important industries in the world. It provides energy for a wide range of applications, including transportation, power generation, and heating. However, the extraction, processing, and transportation of oil and gas also involve significant metallurgical challenges. This chapter will explore the various metallurgical applications in the oil and gas industry and the challenges that must be addressed.

Introduction

The oil and gas industry involves a wide range of metallurgical applications, from drilling and production to refining and transportation. Metallurgy plays a critical role in ensuring the safety and efficiency of these processes. In this chapter, we will explore the various metallurgical challenges associated with the oil and gas industry and the solutions that have been developed to address them.

Corrosion Control

One of the most significant challenges in the oil and gas industry is corrosion. Corrosion can occur in many forms, including general corrosion, localized corrosion, and stress corrosion cracking. It can occur in pipelines, storage tanks, and other equipment used in oil and gas production and transportation. To prevent corrosion, a range of materials and coatings are used, including stainless steel, nickel alloys, and epoxy coatings. The selection of these materials depends on factors such as the environment, temperature, and pressure.

High-Temperature Applications

The oil and gas industry often involves high-temperature applications, such as steam injection and combustion processes. These processes can place significant stresses on the materials used, leading to deformation and failure. To address these challenges, high-temperature alloys such as Inconel, Hastelloy, and titanium are used. These alloys have high strength and corrosion resistance at elevated temperatures.

Welding and Joining

Welding and joining are critical processes in the oil and gas industry, as they are used to join pipelines, storage tanks, and other equipment. However, welding and joining can also lead to metallurgical challenges, such as residual stresses and brittle fracture. To prevent these issues, welding and joining procedures must be carefully controlled, and appropriate filler materials must be used.

Material Selection

The selection of materials is critical in the oil and gas industry, as the materials must be able to withstand harsh environments and conditions. Factors such as temperature, pressure, and the presence of corrosive chemicals must be considered when selecting materials. In addition, materials used in the oil and gas industry must meet stringent regulatory requirements, such as those established by the American Petroleum Institute (API).

Fatigue and Fracture

Fatigue and fracture are also significant challenges in the oil and gas industry, as equipment is subjected to cyclic loading and can experience sudden and catastrophic failure. To prevent fatigue and fracture, materials must be selected based on their fatigue properties and appropriate design and inspection procedures must be implemented.

Non-Destructive Testing

Non-destructive testing (NDT) is critical in the oil and gas industry, as it is used to detect defects and damage in equipment without causing damage. NDT techniques include ultrasonic testing, radiography, magnetic particle testing, and eddy current testing. These techniques are used to detect surface and subsurface defects, cracks, and other types of damage.

Conclusion

Metallurgical applications are critical in the oil and gas industry, as they ensure the safety, reliability, and efficiency of production and transportation processes. The challenges associated with metallurgy in the oil and gas industry require careful consideration of materials, processes, and regulatory requirements. Metallurgists and materials scientists must work closely with engineers and other professionals to address these challenges and ensure the continued success of the industry.

Chapter 27 | The Role of Metallurgy in the Electronics Industry

The electronics industry is a vital sector of the global economy that heavily relies on metallurgy. The manufacture of electronic components and devices involves the use of a wide range of metals and alloys with specific properties and characteristics. This chapter discusses the key role that metallurgy plays in the electronics industry and highlights some of the most common applications of metallurgy in this field.

Overview of the Electronics Industry

The electronics industry encompasses a broad range of businesses and technologies involved in the design, development, manufacturing, and marketing of electronic products and devices. The industry includes manufacturers of semiconductors, printed circuit boards (PCBs), consumer electronics, communication systems, medical equipment, and many other electronic products.

Metallurgical Applications in the Electronics Industry

Metallurgy is an essential component of the electronics industry. Some of the most common applications of metallurgy in this field include:

1. Semiconductors - Semiconductors are at the heart of modern electronics. These materials are made of metals such as silicon, germanium, and gallium arsenide, which are doped with specific impurities to create electrical properties necessary for the functioning of electronic devices.

2. Printed Circuit Boards (PCBs) - PCBs are a critical component of most electronic devices. These boards are made of a non-conductive substrate material, usually fiberglass, that is coated with a layer of copper. Metallurgical processes such as electroplating and etching are used to create the conductive pathways on the PCBs.

3. Connectors and Contacts - Connectors and contacts are used in electronic devices to make electrical connections between different components. These components are often made of metals such as copper, gold, and silver, which have excellent electrical conductivity and resistance to corrosion.

4. Packaging and Enclosures - The packaging and enclosures of electronic devices play a crucial role in protecting the sensitive electronic components from damage and environmental factors. These enclosures are often made of metals such as aluminum, steel, or titanium, which offer high strength and durability.

5. Magnets and Motors - Magnets and motors are used extensively in electronic devices such as hard drives and electric motors. These components often rely on high-performance magnetic materials such as neodymium-iron-boron and samarium-cobalt alloys.

Challenges and Future Directions

As the electronics industry continues to evolve and expand, the demand for new and innovative metallurgical materials and processes will only increase. Metallurgists and materials scientists will need to develop new alloys and manufacturing techniques to meet the specific requirements of the electronics industry, such as miniaturization, high-performance, and low-cost manufacturing. Furthermore, environmental concerns and sustainability issues will also become more prominent in the industry, driving the need for more eco-friendly and sustainable metallurgical processes.

Conclusion

Metallurgy plays a critical role in the electronics industry, providing the materials and processes necessary for the development and manufacturing of electronic devices. The ongoing advancements and innovations in metallurgical materials and processes will continue to shape the future of the electronics industry, enabling the development of smaller, faster, and more powerful electronic devices.

Chapter 28 | Best Practices for Welding Steel: Key Techniques and Considerations

Welding is a crucial process in the fabrication and construction of steel structures, as it involves joining two or more pieces of metal to create a single, unified piece. Welding requires careful attention to detail and adherence to established procedures to ensure high-quality, reliable welds that meet the required specifications. In this chapter, we will discuss some of the best practices for welding steel, including key techniques and considerations that can help improve the quality and efficiency of the welding process.

Welding Techniques:

There are several welding techniques commonly used in steel fabrication, each with its own unique advantages and disadvantages. Some of the most common welding techniques include:

1. Shielded Metal Arc Welding (SMAW): Also known as stick welding, SMAW is a popular welding method that involves using a consumable electrode coated in flux to create the weld. This technique is versatile and can be used on a wide range of steel thicknesses, making it a popular choice for many applications.

2. Gas Metal Arc Welding (GMAW): Also known as MIG welding, GMAW involves using a continuous spool of wire as the electrode, which is fed through a welding gun and melted to create the weld. This method is fast and efficient, making it a popular choice for high-volume welding applications.

3. Flux-Cored Arc Welding (FCAW): Similar to GMAW, FCAW involves using a continuous wire electrode, but the wire is filled with a flux core that provides shielding gas and helps protect the weld from contamination.

Considerations for Welding Steel:

In addition to choosing the right welding technique for the job, there are several other considerations that can help ensure high-quality, reliable welds:

1. Material selection: Choosing the right type of steel for the job is crucial, as different types of steel have different properties that can affect the welding process and the resulting weld. Factors to consider include the thickness of the steel, its chemical composition, and its intended use.

2. Joint preparation: Proper joint preparation is essential to ensure a strong, reliable weld. This includes cleaning the joint of any contaminants, removing any rust or paint, and ensuring the joint is properly aligned and fit together tightly.

3. Welding position: The welding position can affect the quality and strength of the weld. Depending on the position, different welding techniques and settings may be required to achieve the desired results.

4. Welding environment: The welding environment can also affect the quality of the weld. Factors to consider include temperature, humidity, and the presence of any contaminants or gases that could affect the weld.

Conclusion

Welding steel is a complex process that requires careful attention to detail and adherence to established procedures to ensure high-quality, reliable welds. By choosing the right welding technique for the job, considering key factors such as material selection, joint preparation, welding position, and welding environment, and following established best practices and procedures, welders can produce strong, reliable welds that meet the required specifications.

Chapter 29 | The Importance of Refractory Materials in Petrochemical Applications

The petrochemical industry is one of the largest and most important industries in the world. It involves the production of a wide range of chemicals, plastics, and other products from crude oil and natural gas. However, the harsh and extreme conditions in petrochemical plants can cause severe wear and tear on equipment and infrastructure, including the furnaces, reactors, and other high-temperature vessels used in the production process. Refractory materials play a critical role in protecting these structures from heat and corrosion and maintaining their integrity and longevity.

Refractory materials are used in petrochemical applications for several purposes. First, they provide insulation to reduce heat transfer and prevent energy loss, which can help to improve the efficiency and cost-effectiveness of the production process. Second, they act as a barrier between the high-temperature process environment and the metal components of the equipment, preventing corrosion and erosion that can weaken or damage the structure. Finally, they provide support and stability to the equipment, ensuring that it can withstand the stresses and strains of the process.

The selection and installation of refractory materials in petrochemical applications are critical to their performance and durability. Factors such as the type of process, temperature, pressure, and chemical composition of the environment must be carefully considered when choosing the appropriate material. Different types of refractory materials, including bricks, castables, and plastics, have varying properties and strengths that make them suitable for different applications.

In addition to proper material selection, proper installation techniques are also crucial for ensuring the effective performance of refractory materials. Installation considerations such as proper curing time, careful mixing and placement of materials, and proper bonding techniques can all affect the durability and effectiveness of refractory linings.

Conclusion

Overall, the importance of refractory materials in the petrochemical industry cannot be overstated. They play a critical role in protecting equipment and infrastructure from the extreme conditions of high-temperature and corrosive environments, enabling the industry to produce vital chemicals and products efficiently and effectively. Proper material selection and installation techniques are essential for ensuring the long-term performance and durability of these critical components.

Chapter 30 | The Role of Metallurgy in the Defense Industry

Metallurgy plays a crucial role in the defense industry. The military relies on strong, durable, and reliable materials to keep troops safe and equipment operational in harsh and challenging environments. In this chapter, we will explore the various applications of metallurgy in the defense industry, including the development of specialized materials and components for military use.

1. Ballistic Protection: The defense industry relies heavily on metallurgy to create materials that can withstand the impact of high-velocity projectiles. These materials include armor plate, armor-piercing projectiles, and armor-penetrating materials, which are used to protect military vehicles and personnel from enemy fire.

2. Aircraft Materials: The aerospace industry has driven the development of many high-performance alloys, which have since found application in the defense industry. These alloys provide excellent strength-to-weight ratios, high temperature resistance, and resistance to corrosion, making them ideal for use in aircraft components such as jet engine blades, landing gear, and wing structures.

3. Missile Components: Missiles are complex systems that rely on a variety of metallurgical components, including guidance systems, fuel tanks, and propulsion systems. These components must be able to withstand extreme temperatures, high pressures, and corrosive environments to operate effectively.

4. Communications and Electronics: The defense industry relies on sophisticated communications and electronics systems to coordinate operations and gather intelligence. Metallurgy plays a key role in the development of these systems, as it is used to create specialized materials that can handle high-frequency electromagnetic waves and resist corrosion from exposure to the environment.

5. Explosive Devices: The defense industry uses metallurgical expertise to create specialized materials for explosive devices such as bombs and landmines. These materials must be durable, stable, and sensitive to detonation in order to be effective.

6. Vehicle Armor: Military vehicles must be protected from enemy fire, and metallurgical expertise is used to create specialized armor plates that can provide this protection. These materials must be strong, durable, and lightweight, as they must be able to protect the vehicle without hindering its mobility.

7. Specialized Components: The defense industry relies on a wide range of specialized components, from aircraft landing gear to satellite components, that require unique metallurgical properties. These components must be designed to withstand extreme temperatures, high pressures, and corrosive environments, while also meeting strict size and weight requirements.

Conclusion

Overall, the role of metallurgy in the defense industry is crucial to ensuring the safety and effectiveness of military operations. The development of specialized materials and components is key to keeping troops safe and equipment operational in challenging environments.

Chapter 31 | A Guide to Understanding Steel Microstructures

Steel is a widely used material in many industries due to its high strength, durability, and versatility. The properties of steel, including its strength, toughness, and corrosion resistance, depend largely on its microstructure. The microstructure of steel refers to its crystal structure and the distribution of its different constituents at a microscopic level. Understanding the microstructure of steel is critical for its proper selection, processing, and application.

The microstructure of steel is influenced by many factors, including the chemical composition of the steel, the heat treatment it undergoes, and the processing conditions during its manufacture. The microstructure of steel can be observed and analyzed using different techniques, such as optical microscopy, scanning electron microscopy, transmission electron microscopy, X-ray diffraction, and electron backscatter diffraction.

Steel microstructures can be broadly classified into three categories: ferrite, pearlite, and martensite. Ferrite is a soft, ductile, and relatively weak microstructure, consisting of iron atoms arranged in a body-centered cubic crystal structure. Pearlite is a mixture of ferrite and cementite, a hard and brittle compound of iron and carbon, arranged in a layered pattern. Martensite is a hard and brittle microstructure that forms when austenite, a high-temperature phase of iron, is rapidly cooled to room temperature.

Other microstructures that can be found in steel include bainite, a mixture of ferrite and cementite that forms when austenite is cooled at an intermediate rate, and tempered martensite, which results from the reheating of martensite to a lower temperature. The microstructure of steel can be modified by controlling its cooling rate, heating rate, and annealing temperature.

In addition to its microstructure, steel properties can be influenced by the presence of different alloying elements. Alloying elements such as chromium, molybdenum, nickel, and vanadium can improve the corrosion resistance, strength, and toughness of steel.

Understanding the microstructure of steel is critical for its successful application in different industries. Proper selection of steel with the right microstructure can ensure its optimum performance in various applications. The development of new steel microstructures and alloy compositions is also critical for the advancement of steel technology and its applications in emerging industries.

CHAPTER 32 | THE IMPORTANCE OF METALLURGY IN BUILDING AND CONSTRUCTION

Metallurgy plays a critical role in the building and construction industry by providing a wide range of materials with unique properties that are essential for building structures that are safe, efficient, and long-lasting. This chapter will provide an overview of the role of metallurgy in building and construction, and will cover the different types of metals and alloys used in this industry, their properties, and applications.

Types of Metals and Alloys Used in Building and Construction

The building and construction industry uses a variety of metals and alloys, including steel, aluminum, copper, titanium, and zinc. Each of these materials has its unique properties that make them suitable for specific applications. Steel, for example, is used extensively in building construction due to its high strength, durability, and cost-effectiveness. Aluminum is commonly used in windows and doors due to its lightweight, corrosion-resistant properties. Copper is used in roofing materials due to its excellent thermal and electrical conductivity. Titanium is used in high-performance structures due to its high strength-to-weight ratio, while zinc is commonly used in galvanized coatings to provide corrosion resistance.

Properties and Applications

The properties of metals and alloys used in building and construction vary widely, and their selection depends on the intended use of the material. For example, steel is commonly used in building construction due to its high strength, ductility, and durability. It can be fabricated into various shapes and sizes and can be easily welded, making it suitable for a wide range of applications, including beams, columns, and framing. Aluminum, on the other hand, is lightweight and corrosion-resistant, making it suitable for use in windows, doors, and cladding. Copper is highly conductive, making it suitable for electrical applications, while zinc is highly corrosion-resistant, making it ideal for use in roofing materials.

Advances in Metallurgy

Advances in metallurgy have led to the development of new materials with unique properties that offer superior performance in building and construction applications. For example, advanced high-strength steels (AHSS) have been developed to provide increased strength and durability, while reducing weight and material costs. Similarly, the development of shape memory alloys (SMAs) has led to the creation of materials that can change shape in response to external stimuli, making them ideal for use in seismic-resistant structures.

Conclusion

Metallurgy plays a critical role in the building and construction industry, providing a range of materials with unique properties that are essential for building structures that are safe, efficient, and long-lasting. As the industry continues to evolve, the development of new materials and advances in metallurgical technologies will continue to drive innovation in building design and construction.

CHAPTER 33 | THE ROLE OF METALLURGY IN THE FOOD AND BEVERAGE INDUSTRY

Metallurgy may not be the first thing that comes to mind when thinking about the food and beverage industry, but it plays a crucial role in ensuring the safety and quality of food products. Metallurgical materials and processes are used to manufacture equipment, such as stainless steel tanks, pipes, and fittings, which are used to handle, store, and process food and beverages. This chapter will discuss the importance of metallurgy in the food and beverage industry and the various applications of metallurgical materials and processes in this sector.

Metallurgical Materials Used in the Food and Beverage Industry

The food and beverage industry requires materials that are resistant to corrosion, oxidation, and bacterial growth. Stainless steel is a common choice for equipment used in this industry due to its excellent resistance to corrosion, ease of cleaning, and durability. Other metallurgical materials, such as titanium and nickel alloys, are also used in the food and beverage industry for specialized applications.

Metallurgical Processes Used in the Food and Beverage Industry

Metallurgical processes are used to fabricate, shape, and join metals into the desired equipment for the food and beverage industry. Processes such as welding, brazing, and soldering are used to join metal components, while machining and forming processes are used to shape the equipment. Coatings and surface treatments are also applied to the equipment to improve its properties, such as corrosion resistance and wear resistance.

Applications of Metallurgy in the Food and Beverage Industry

Metallurgy plays a critical role in the various stages of food and beverage production, from processing and storage to packaging and transportation. Stainless steel equipment is used to handle and store food products, such as dairy, meat, and beverages. Metal detectors are used to detect and remove metal contaminants in food products, while food-grade coatings are used to protect food packaging materials from contamination.

Conclusion

Metallurgy is a vital aspect of the food and beverage industry, ensuring that food products are handled, stored, and processed safely and efficiently. The use of metallurgical materials and processes, such as stainless steel and specialized coatings, contributes to the quality and safety of food products. As the food and beverage industry continues to evolve, advancements in metallurgical technology will play an essential role in developing new and innovative solutions to meet the industry's evolving needs.

Chapter 34 | The Advantages and Disadvantages of Different Casting Techniques

Casting is a widely used manufacturing process in the metallurgical industry that involves the pouring of molten metal into a mold cavity to produce a solidified part. There are different casting techniques available, and each has its own advantages and disadvantages. In this chapter, we will explore some of the most common casting techniques and their strengths and limitations.

1. Sand Casting

Sand casting is one of the oldest and most widely used casting techniques. In this method, a pattern made of wood or other materials is pressed into a sand mixture to form a mold cavity. Molten metal is then poured into the mold and allowed to cool and solidify. Sand casting has the advantage of being relatively inexpensive and able to produce large, complex parts. However, the surface finish of sand castings may be rough, and the process can be time-consuming.

2. Investment Casting

Investment casting, also known as lost-wax casting, is a precision casting process that involves the creation of a wax pattern. The pattern is coated with a ceramic material, which hardens and creates a mold. The wax is melted out of the mold, and molten metal is poured in. Investment casting can produce intricate shapes with excellent surface finishes, but it is typically more expensive than sand casting.

3. Die Casting

Die casting is a process in which molten metal is injected under high pressure into a steel mold, known as a die. The mold is designed to produce a specific shape, and the high pressure helps to ensure that the part is formed accurately. Die casting is ideal for producing high volumes of small to medium-sized parts with excellent dimensional accuracy. However, die casting can be expensive due to the cost of the dies and requires significant up-front investment.

4. Continuous Casting

Continuous casting is a process used to cast metals into long, continuous shapes, such as bars, rods, or tubes. In this process, molten metal is poured into a mold and cooled using water or other cooling media. The casting is then pulled out continuously, allowing for a long, continuous piece of metal to be formed. Continuous casting has the advantage of producing high-quality, defect-free metal with consistent dimensions. However, the process is only suitable for certain types of metals and requires specialized equipment.

5. Centrifugal Casting

Centrifugal casting is a process that uses centrifugal force to distribute molten metal into a mold. The mold is rotated at high speed, and the metal is poured into the mold while it is spinning. The centrifugal force helps to distribute the molten metal evenly, resulting in parts with excellent mechanical properties. Centrifugal casting is ideal for producing cylindrical parts, such as pipes and tubes, but can also produce parts with complex shapes. However, the process is limited to certain types of metals and can be difficult to control.

6. Vacuum Casting

Vacuum casting is a process in which a vacuum is used to remove air from the mold cavity before the molten metal is poured in. This process helps to reduce porosity in the final part and improve its mechanical properties. Vacuum casting is ideal for producing parts with high dimensional accuracy and excellent surface finish. However, the process can be expensive due to the need for specialized equipment and can be limited to certain types of metals.

Conclusion

In conclusion, each casting technique has its own set of advantages and disadvantages. The choice of which technique to use depends on factors such as the desired part geometry, material properties, and production volume. A thorough understanding of the different casting techniques and their limitations is crucial to selecting the most appropriate method for a particular application.

Chapter 35 | The Role of Metallurgy in the Mining Industry

Metallurgy plays a vital role in the mining industry, where metals and minerals are extracted from the earth for various purposes. Metallurgical engineers use a variety of techniques to extract, process, and refine metals and minerals. They work with a range of metals, including copper, gold, silver, lead, zinc, iron, and nickel, as well as non-metallic minerals such as coal, diamonds, and uranium.

The primary goal of metallurgical engineers in the mining industry is to extract metals and minerals from the ground in the most efficient and cost-effective manner possible. This involves a range of processes, including exploration, mine planning, extraction, processing, refining, and waste management.

Exploration involves identifying the location and quality of mineral deposits in the ground. Metallurgical engineers work with geologists and other mining professionals to conduct geological surveys, drilling, and sampling to assess the size and composition of the deposits.

Once the mineral deposits have been identified, mining engineers use a variety of techniques to extract them from the ground. These can include underground mining, open-pit mining, or a combination of both. Metallurgical engineers must take into account the properties of the minerals being mined, as well as the local environmental conditions, when determining the most appropriate extraction method.

Once the minerals have been extracted from the ground, they need to be processed and refined to remove impurities and increase their purity. This involves a range of metallurgical techniques, including crushing, grinding, flotation, smelting, and refining. Metallurgical engineers work closely with chemical engineers, process engineers, and other mining professionals to develop and optimize these processes.

In addition to extracting and refining metals and minerals, metallurgical engineers in the mining industry are also responsible for managing the waste products produced during the mining and refining processes. This can include tailings, waste rock, and other byproducts. Metallurgical engineers work to develop and implement environmentally sustainable waste management practices to minimize the impact of mining activities on the surrounding environment.

Conclusion

Overall, metallurgical engineers play a critical role in the mining industry by ensuring the efficient and responsible extraction, processing, and refining of metals and minerals. They work closely with other mining professionals to develop and implement innovative and sustainable mining practices that meet the growing demand for minerals and metals around the world.

Chapter 36 | Heat Exchanger Design and Optimization in Metallurgical Applications

Heat exchangers are essential components in many metallurgical processes, as they are used to transfer heat from one medium to another. These devices are used in a wide range of applications, including the cooling and heating of liquids, gases, and solids. Heat exchangers are used in metallurgical applications to control temperatures during various stages of production, including melting, refining, and casting.

Heat exchangers can be designed and optimized for specific metallurgical applications, and the selection of the right heat exchanger is critical to ensure efficient and effective heat transfer. This chapter will discuss the various types of heat exchangers used in metallurgical applications, design considerations, and optimization techniques.

Types of Heat Exchangers

There are several types of heat exchangers used in metallurgical applications, including:

1. Shell and Tube Heat Exchangers: This is the most common type of heat exchanger used in metallurgical applications. In this design, one fluid flows through tubes while the other flows through a shell around the tubes. This type of heat exchanger is highly efficient and can handle high pressure and temperature differentials.

2. Plate Heat Exchangers: This type of heat exchanger uses plates to transfer heat between fluids. The plates are arranged in a stack and have channels for the flow of fluids. Plate heat exchangers are highly efficient and are ideal for applications where space is limited.

3. Spiral Heat Exchangers: This type of heat exchanger uses a spiral design to transfer heat between fluids. The design of the spiral creates a high degree of turbulence, which increases heat transfer efficiency. Spiral heat exchangers are highly efficient and are ideal for applications where a high degree of turbulence is required.

Design Considerations

When designing a heat exchanger for metallurgical applications, several factors need to be considered, including:

1. Fluid Properties: The properties of the fluids being used, including viscosity, thermal conductivity, and density, are essential in determining the type of heat exchanger to use.

2. Operating Conditions: The operating conditions, including temperature and pressure, must be considered when designing a heat exchanger. The design must be capable of handling the expected pressure and temperature differentials.

3. Material Selection: The material selection is critical in ensuring that the heat exchanger can withstand the corrosive environment of metallurgical processes.

Optimization Techniques

Several techniques can be used to optimize heat exchangers in metallurgical applications, including:

1. Increasing Flow Velocity: Increasing the flow velocity of fluids through a heat exchanger can improve heat transfer efficiency.

2. Increasing Surface Area: Increasing the surface area of the heat exchanger can also improve heat transfer efficiency. This can be achieved by using a larger heat exchanger or increasing the number of plates or tubes.

3. Reducing Fouling: Fouling occurs when contaminants accumulate on the heat transfer surface, reducing efficiency. Fouling can be minimized by selecting the right materials and cleaning the heat exchanger regularly.

Conclusion

Heat exchangers are essential components in metallurgical applications, and selecting the right heat exchanger is critical to ensuring efficient and effective heat transfer. The design of the heat exchanger must consider the fluid properties, operating conditions, and material selection to ensure that it can handle the expected pressure and temperature differentials. Optimization techniques such as increasing flow velocity, increasing surface area, and reducing fouling can be used to improve heat transfer efficiency.

CHAPTER 37 | BEST PRACTICES FOR CORROSION TESTING AND PREVENTION

Corrosion is a natural process that occurs in metals and alloys when exposed to various environmental conditions. The degradation of metal components through corrosion can have severe consequences, leading to costly maintenance and repair, or even catastrophic failures. Therefore, it is essential to implement effective corrosion testing and prevention practices to ensure the longevity and reliability of metal components.

This chapter provides an overview of the best practices for corrosion testing and prevention, including the importance of selecting appropriate testing methods and materials, establishing a comprehensive corrosion prevention program, and adopting the latest corrosion-resistant technologies.

Corrosion Testing Methods

Corrosion testing is an essential tool to evaluate the corrosion resistance of metals and alloys. The selection of an appropriate testing method depends on the specific application, materials, and environmental conditions.

The most commonly used corrosion testing methods include:

1. Salt Spray Testing

Salt spray testing is a widely used corrosion testing method that involves exposing the test specimen to a highly corrosive environment of saltwater mist. This method can simulate harsh coastal environments and provide a quick and easy evaluation of a material's corrosion resistance.

2. Electrochemical Testing

Electrochemical testing is a non-destructive testing method that involves measuring the corrosion potential and current of a metal in a corrosive environment. This method can provide valuable information on the corrosion resistance of the material and the rate of corrosion.

3. Immersion Testing

Immersion testing involves immersing the test specimen in a corrosive solution for a specified period. This method can provide a more realistic evaluation of a material's corrosion resistance in actual operating conditions.

Corrosion Prevention

Corrosion prevention is an ongoing process that involves identifying potential corrosion risks and implementing appropriate preventive measures. The most effective corrosion prevention programs include a combination of design, material selection, and protective coatings.

1. Design

Design considerations play a crucial role in preventing corrosion. The use of appropriate materials, geometries, and surface finishes can help minimize the impact of corrosive environments on metal components. For instance, designing components with adequate drainage and ventilation can prevent water accumulation and promote natural drying, reducing the risk of corrosion.

2. Material Selection

Choosing the right material for a specific application is critical to prevent corrosion. Material selection should consider the environmental conditions, including temperature, humidity, and the presence of corrosive agents such as acids or salts. Corrosion-resistant materials, such as stainless steel or titanium, are often preferred for harsh environments.

3. Protective Coatings

Protective coatings, such as paints, varnishes, or electroplated coatings, can provide an additional layer of protection to metal components. Coatings act as a barrier between the metal surface and the corrosive environment, preventing or delaying the onset of corrosion. However, proper surface preparation and application procedures are crucial to ensure the effectiveness of coatings.

Emerging Technologies for Corrosion Prevention

In recent years, several innovative technologies have been developed to prevent corrosion. These technologies include:

1. Corrosion Inhibitors

Corrosion inhibitors are chemical compounds that can reduce the rate of corrosion by adsorbing onto the metal surface and forming a protective film. Inhibitors can be added to the corrosive environment or applied as a coating to the metal surface.

2. Cathodic Protection

Cathodic protection involves the use of a sacrificial anode or an external power source to supply electrons to the metal surface, creating a cathodic zone that prevents corrosion. This method is often used to protect pipelines and other submerged metal structures.

3. Self-healing coatings

These coatings contain microcapsules filled with healing agents that can be released upon damage, restoring the coating's protective properties.

4. Nanocoatings

Nanocoatings are extremely thin coatings that provide a high level of protection against corrosion due to their high surface area and the ability to be tailored to specific metals and environments.

5. Electrochemical techniques

Electrochemical techniques, such as impressed current cathodic protection and electrochemical impedance spectroscopy, can provide active protection against corrosion by applying a current or measuring the electrochemical behavior of the metal.

6. Atomic layer deposition

Atomic layer deposition is a technique that can deposit extremely thin layers of materials with precise control, allowing for the creation of corrosion-resistant coatings with excellent adhesion and coverage.

7. Plasma-based surface modification

Plasma-based surface modification can alter the surface chemistry and properties of metals, providing enhanced corrosion resistance and adhesion.

8. Biocorrosion prevention

Biocorrosion is a type of corrosion caused by microorganisms, and emerging technologies are being developed to prevent and mitigate biocorrosion by utilizing natural antimicrobial agents or engineering surfaces to resist microbial adhesion.

As with any emerging technology, further research and testing are needed to fully understand and optimize these corrosion prevention methods. However, they offer promising solutions for reducing the economic and safety impact of corrosion on various industries.

Monitoring and Maintenance

Finally, it is important to monitor the condition of the metal surface and perform regular maintenance to prevent corrosion. This includes visual inspection, non-destructive testing, and cleaning. Visual inspection can help identify any signs of corrosion, such as discoloration, pitting, or rusting. Non-destructive testing, such as ultrasonic testing and radiography, can detect corrosion that may not be visible to the naked eye. Cleaning the metal surface regularly can remove any contaminants that may accelerate corrosion.

Conclusion

Corrosion is a serious problem that can have significant economic and safety implications. However, by understanding the factors that contribute to corrosion and implementing effective prevention and control measures, it is possible to minimize the impact of corrosion and extend the lifespan of metal components. By following the best practices outlined in this chapter, metallurgical professionals can help ensure that their products and structures are protected from corrosion and maintain their integrity over time.

CHAPTER 38 | THE ROLE OF METALLURGY IN BIOMEDICAL APPLICATIONS

Metallurgy plays a vital role in the development of biomedical devices, implants, and other medical applications. The selection of appropriate metals and alloys, their processing, and surface modification techniques are critical factors that determine the success of these biomedical applications. In this chapter, we will discuss the role of metallurgy in various biomedical applications, including implant materials, surgical instruments, and diagnostic tools.

Implant Materials:

Metals and alloys are widely used as implant materials due to their excellent mechanical properties, biocompatibility, and corrosion resistance. The most commonly used metals and alloys include titanium, stainless steel, cobalt-chromium alloys, and nickel-titanium alloys.

Titanium and its alloys are widely used in orthopedic and dental implants due to their excellent biocompatibility, low modulus of elasticity, and corrosion resistance. Stainless steel is used in cardiovascular stents and bone fixation devices due to its high strength and excellent fatigue properties. Cobalt-chromium alloys are used in hip and knee implants due to their high strength, wear resistance, and biocompatibility. Nickel-titanium alloys are used in dental and orthopedic implants due to their shape memory and superelasticity properties.

Surgical Instruments:

Metals and alloys are also used in the manufacture of surgical instruments. Stainless steel is the most commonly used material due to its excellent corrosion resistance, high strength, and ease of fabrication. Other materials such as titanium, cobalt-chromium alloys, and nickel-titanium alloys are also used in specialized surgical instruments.

Diagnostic Tools:

Metals and alloys are also used in the manufacture of diagnostic tools such as X-ray tubes, CT scanners, and MRI machines. Tungsten and molybdenum alloys are used in X-ray tubes due to their high melting points and high thermal conductivity. Cobalt-chromium alloys are used in CT scanners due to their high strength and corrosion resistance. MRI machines use superconducting magnets made of niobium-titanium alloys.

Surface Modification Techniques:

Surface modification techniques such as surface coatings, surface treatments, and surface texturing are used to improve the biocompatibility and wear resistance of implant materials. Various surface coating techniques such as plasma spraying, electrophoretic deposition, and physical vapor deposition are used to deposit biocompatible materials such as hydroxyapatite, titanium nitride, and diamond-like carbon onto implant surfaces.

Conclusion:

In conclusion, metallurgy plays a critical role in the development of biomedical applications. The selection of appropriate metals and alloys, their processing, and surface modification techniques are critical factors that determine the success of these biomedical applications. With advancements in materials science and engineering, new and improved biomaterials and implant designs are continuously being developed to improve the performance and longevity of medical devices and implants.

CHAPTER 39 | THE FUTURE OF STEEL: ADVANCEMENTS AND INNOVATIONS

Steel has been a crucial material for human civilization for centuries, used in everything from construction and transportation to medical devices and weaponry. As technology has advanced, so too has the production and use of steel. The future of steel promises even greater advancements and innovations, as new technologies and materials are developed to meet the needs of our ever-evolving world.

In this chapter, we will explore the current state of the steel industry, the challenges it faces, and the exciting advancements being made in steel production and application.

Current State of the Steel Industry:

The steel industry is a massive global market, with an estimated value of $2.5 trillion in 2020. Steel production continues to grow, with over 1.8 billion metric tons produced worldwide in 2020. China is the largest producer of steel, accounting for around 60% of global production, followed by India, Japan, and the United States.

However, the steel industry faces several challenges, including increasing competition, environmental regulations, and shifting consumer demands. In response, the industry is exploring new materials, technologies, and production methods to improve efficiency and sustainability.

Advancements and Innovations:

1. Green Steel: One of the most significant advancements in steel production is the development of "green steel." This refers to steel that is produced using renewable energy and low-carbon technologies, reducing greenhouse gas emissions and environmental impact. This includes using hydrogen-based reduction processes, electric arc furnaces powered by renewable energy, and carbon capture and storage technologies.

2. Smart Steel: Another exciting innovation in the steel industry is the development of "smart steel." This refers to steel that can sense and respond to its environment, providing valuable data on structural integrity, temperature, and stress. Smart steel can be used in a variety of applications, including bridges, buildings, and pipelines, to monitor performance and detect potential problems before they become catastrophic.

3. Advanced Alloys: Metallurgists are developing new alloys with improved properties, including strength, ductility, and corrosion resistance. This includes high-strength steels for automotive and aerospace applications, corrosion-resistant alloys for marine environments, and shape-memory alloys for medical devices.

4. Additive Manufacturing: Additive manufacturing, or 3D printing, is being explored as a potential method for producing steel components with complex geometries and customized properties. This technology allows for greater design flexibility and reduced waste, potentially revolutionizing the way steel is produced and used.

5. Nanotechnology: Nanotechnology is being used to develop new coatings and surface treatments for steel, improving its resistance to wear, corrosion, and oxidation. This includes using nanoparticles to create self-healing coatings and surface textures that reduce friction and improve energy efficiency.

Conclusion:

The future of steel is bright, with exciting advancements and innovations being made in production and application. The development of green steel, smart steel, advanced alloys, additive manufacturing, and nanotechnology are just a few of the areas where metallurgists and engineers are pushing the boundaries of what is possible with this versatile and essential material. As the world continues to evolve, the steel industry will continue to play a vital role in meeting the needs of our changing society.

Chapter 40 | Understanding the Role of Refractory Materials in Aluminum Manufacturing

Aluminum is a widely used metal in many industrial applications, including construction, transportation, and packaging. The manufacturing process of aluminum involves several stages, including refining, smelting, casting, and finishing. The high temperatures and harsh chemical environments present in these processes make refractory materials an essential component of the manufacturing infrastructure. This chapter aims to explore the critical role of refractory materials in aluminum manufacturing, including their properties, types, and applications.

Properties of Refractory Materials

Refractory materials are designed to withstand high temperatures, corrosion, and abrasion. They are typically characterized by their refractoriness, thermal shock resistance, and mechanical strength. The most common refractory materials used in aluminum manufacturing include alumina, magnesia, and silica. Alumina is the most widely used refractory material due to its high melting point, excellent thermal shock resistance, and chemical inertness. Magnesia and silica are also commonly used due to their high refractoriness and corrosion resistance.

Types of Refractory Materials

There are two main types of refractory materials used in aluminum manufacturing: bricks and monolithics. Refractory bricks are precast, rectangular blocks made of refractory material that are used to line furnaces, kilns, and other high-temperature equipment. Monolithic refractories, on the other hand, are unshaped refractory materials that can be cast in place or sprayed onto a surface. Monolithic refractories are typically used to repair or replace worn-out refractory linings in high-temperature equipment.

Applications of Refractory Materials in Aluminum Manufacturing

Refractory materials are used in several stages of aluminum manufacturing. In the refining stage, refractory materials are used to line furnaces used to smelt alumina into aluminum. In the smelting stage, refractory materials are used to line electrolytic cells, which are used to extract aluminum from alumina. In the casting stage, refractory materials are used to line molds used to cast aluminum into various shapes. In the finishing stage, refractory materials are used to line furnaces used to anneal and temper aluminum.

Challenges in Refractory Materials Selection and Performance

The selection and performance of refractory materials in aluminum manufacturing are influenced by several factors, including the chemical composition of the molten aluminum, the temperature and pressure of the process, and the abrasiveness of the materials being processed. Refractory materials can experience wear and tear due to thermal shock, mechanical stress, and chemical attack, which can lead to refractory failure and production downtime.

Future Directions in Refractory Materials for Aluminum Manufacturing

The future of refractory materials in aluminum manufacturing lies in the development of new materials with higher refractoriness, better thermal shock resistance, and improved corrosion resistance. Additionally, there is a growing interest in the use of advanced ceramic materials, such as silicon carbide and zirconia, in high-temperature applications. The use of robotics and automation in refractory installation and maintenance is also expected to increase in the future, leading to more efficient and cost-effective manufacturing processes.

Conclusion

Refractory materials play a critical role in aluminum manufacturing, providing essential protection against high temperatures, corrosion, and abrasion. Understanding the properties, types, and applications of refractory materials is essential to ensure the safe and efficient operation of aluminum manufacturing processes. Advances in refractory materials technology are expected to drive continued improvements in aluminum manufacturing efficiency, cost-effectiveness, and sustainability.

Chapter 41 | The Importance of Metallurgy in the Recycling Industry

The recycling industry plays an important role in reducing waste and conserving natural resources. Metallurgy also plays a significant role in the recycling industry, as it is essential for the recovery and reuse of valuable metals. This chapter aims to highlight the importance of metallurgy in the recycling industry, including the challenges and opportunities associated with the recycling of metals.

Metals Recycling and Metallurgical Processes:

Metals are widely used in various industries, such as automotive, aerospace, construction, and electronics. However, the production of metals requires significant amounts of energy and resources. Recycling metals is an effective way to reduce waste and conserve resources. Metallurgical processes are used to recover metals from scrap materials, such as old cars, appliances, and electronic devices. The metallurgical processes include sorting, crushing, melting, refining, and casting. These processes require specialized knowledge and equipment to ensure the quality and purity of the recovered metals.

Challenges in the Recycling Industry:

The recycling industry faces several challenges in the recovery and reuse of metals. One of the main challenges is the quality of the scrap materials. Scrap materials can be contaminated with other materials, such as plastics and glass, which can affect the quality of the recovered metals. Another challenge is the high cost of recycling. Metallurgical processes can be energy-intensive, and the cost of energy can significantly affect the profitability of recycling operations. Additionally, the demand for recycled metals can fluctuate, making it difficult for recycling companies to predict market demand.

Opportunities in the Recycling Industry:

Despite the challenges, the recycling industry presents several opportunities for metallurgy. The growing demand for sustainable materials and the circular economy presents opportunities for the recycling industry to expand. Additionally, advancements in metallurgical processes and technologies can improve the efficiency and cost-effectiveness of recycling operations. For example, new techniques such as plasma processing and electrochemical processes can be used to recover metals from scrap materials more efficiently and at a lower cost. Furthermore, the use of recycled metals can reduce the environmental impact of metal production, as it requires less energy and resources compared to primary metal production.

Conclusion:

The importance of metallurgy in the recycling industry cannot be overstated. The recovery and reuse of valuable metals not only reduce waste and conserve resources but also provide economic and environmental benefits. However, the recycling industry faces several challenges, such as the quality of scrap materials and the high cost of recycling operations. To address these challenges, the industry must continue to invest in research and development to improve the efficiency and cost-effectiveness of metallurgical processes. Overall, metallurgy plays a critical role in the recycling industry and will continue to do so as the demand for sustainable materials and the circular economy grows.

Chapter 42 | Quality Assurance and Control in Steelmaking

Quality assurance and control are essential in any manufacturing process, especially in steelmaking. The steel industry has evolved over the years, and the demand for high-quality steel products has increased. Steel manufacturers have to ensure that the products they produce meet the required quality standards. This chapter will explore the concept of quality assurance and control in steelmaking, its importance, and the techniques used to achieve it.

Importance of Quality Assurance and Control in Steelmaking:

Quality assurance and control are crucial in steelmaking for several reasons. Firstly, it ensures that the end product meets the required standards, which leads to customer satisfaction. Secondly, it helps to reduce waste and increase productivity, which results in cost savings. Thirdly, it ensures that the steel produced is safe to use and does not pose any hazards to consumers or the environment. Finally, it helps to build the reputation of the manufacturer and enhances their brand image.

Quality Assurance Techniques in Steelmaking:

Quality assurance in steelmaking involves several techniques, which are used to monitor the process and ensure that the end product meets the required standards. Some of the techniques used include:

1. Statistical Process Control (SPC): This technique involves monitoring the process variables and using statistical methods to detect any deviations from the set standards. SPC helps to identify any process variations and enables manufacturers to take corrective action before the product is produced.

2. Six Sigma: This is a quality control technique that aims to reduce defects in the manufacturing process. Six Sigma uses statistical analysis to identify and eliminate the root causes of defects.

3. Total Quality Management (TQM): TQM is a management philosophy that focuses on continuous improvement in all aspects of the manufacturing process. TQM involves the participation of all employees in the process, and their input is used to improve the process continually.

4. Failure Mode and Effects Analysis (FMEA): FMEA is a technique used to identify and analyze potential failure modes in the manufacturing process. The technique involves identifying potential failure modes, analyzing their effects, and taking corrective action to prevent them.

5. Control Charts: Control charts are used to monitor the process variables over time. The charts show the upper and lower control limits, and any variation from these limits can be detected and corrected.

Conclusion:

Quality assurance and control are essential in steelmaking to ensure that the end product meets the required quality standards. The techniques used to achieve this include statistical process control, Six Sigma, total quality management, failure mode and effects analysis, and control charts. The implementation of quality assurance and control techniques helps to reduce waste, increase productivity, and enhance the reputation of the manufacturer.

Chapter 43 | The Role of Metallurgy in the Water Treatment Industry

Metallurgy plays an essential role in the water treatment industry by providing the necessary materials and technologies to treat, purify, and distribute clean water to communities. The demand for clean water is increasing, and metallurgical applications are continuously evolving to meet this need. This chapter will explore the different applications of metallurgy in the water treatment industry and its role in providing safe and clean water to communities.

Materials Used in Water Treatment:

Metallurgical applications provide a wide range of materials that are used in water treatment processes. One of the most common materials is stainless steel, which is widely used in water treatment plants due to its resistance to corrosion and its ability to withstand high temperatures and pressures. Other materials used in water treatment processes include nickel alloys, titanium, and copper alloys.

Metallurgical Applications in Water Treatment:

Metallurgy plays a crucial role in the water treatment industry by providing various technologies and applications. One of the most significant applications is the use of membranes in water treatment processes. Membranes are used to remove impurities, bacteria, and viruses from water, and metallurgical technologies play an essential role in producing membranes that are resistant to corrosion, fouling, and scaling.

Another significant application of metallurgy in water treatment is the production of pumps and valves used in water distribution and treatment systems. These pumps and valves must be resistant to corrosion and erosion, and metallurgical technologies provide materials that can withstand harsh environments and provide long-lasting performance.

Metallurgical applications also provide coatings and linings that are used in water treatment processes. These coatings and linings provide protection against corrosion and ensure the longevity of equipment used in water treatment processes.

Challenges in Metallurgy for Water Treatment:

One of the significant challenges in metallurgy for the water treatment industry is the need to produce materials that can withstand high temperatures, pressures, and harsh chemical environments. The materials used in water treatment processes must be resistant to corrosion, erosion, and fouling, which can be challenging to achieve in some applications.

Another challenge in metallurgy for water treatment is the need to produce materials that are cost-effective while providing long-lasting performance. This requires a balance between material performance and cost, which can be challenging to achieve.

Conclusion:

Metallurgy plays an essential role in the water treatment industry by providing the necessary materials and technologies to treat, purify, and distribute clean water to communities. The use of stainless steel, nickel alloys, titanium, and copper alloys in water treatment processes is vital in producing equipment that is resistant to corrosion, erosion, and fouling. With the increasing demand for clean water, metallurgical applications are continuously evolving to meet the needs of the water treatment industry.

CHAPTER 44 | BEST PRACTICES FOR STEEL HEAT TREATMENT PROCESSES

Heat treatment is a critical process in the manufacturing of steel components. It is the process of heating and cooling steel to modify its physical and mechanical properties, such as hardness, strength, toughness, and ductility. The heat treatment process consists of three stages: heating, soaking, and cooling. The type of heat treatment process used depends on the intended application of the steel component. In this chapter, we will discuss best practices for steel heat treatment processes.

1. Understanding Steel Composition:
Steel composition plays a crucial role in determining the heat treatment process required. Different grades of steel have varying compositions and respond differently to heat treatment processes. It is essential to understand the chemical composition of the steel before selecting the appropriate heat treatment process.

2. Proper Quenching Techniques:
Quenching is the process of cooling heated steel rapidly to room temperature to harden it. The choice of quenching medium, such as oil, water, or air, depends on the steel composition and desired properties. Proper quenching techniques must be followed to avoid cracking or warping of the steel component.

3. Precise Temperature Control:
Precise temperature control is critical in the heat treatment process. The temperature must be maintained within the specified range for the required duration to achieve the desired properties. Failure to control temperature within the specified range can result in poor quality components.

4. Understanding Cooling Rates:
Different heat treatment processes require different cooling rates to

achieve the desired properties. It is crucial to understand the required cooling rate for the specific process and ensure that it is achieved during the cooling stage.

5. Proper Maintenance of Heat Treatment Equipment:

Heat treatment equipment such as furnaces, quenching tanks, and cooling systems require regular maintenance to ensure their proper functioning. Failure to maintain these components can result in poor quality components and equipment failure.

6. Inspection and Quality Control:

Inspection and quality control are essential to ensure that the heat treatment process has achieved the desired results. The quality control process should include non-destructive testing, such as ultrasonic testing, to identify any defects or irregularities in the steel component.

Conclusion:

The steel heat treatment process is critical to the manufacturing of high-quality steel components. Proper understanding of steel composition, precise temperature control, proper quenching techniques, understanding cooling rates, proper maintenance of heat treatment equipment, and inspection and quality control are key best practices for achieving the desired properties in steel components. By following these best practices, manufacturers can produce high-quality steel components that meet the required specifications and standards.

CHAPTER 45 | THE IMPORTANCE OF MATERIAL CHARACTERIZATION IN METALLURGY

Metallurgy is a branch of material science that deals with the physical and chemical behavior of metals and alloys. Material characterization is an essential aspect of metallurgy, which involves analyzing the material's microstructure, mechanical, physical, and chemical properties. The knowledge of material characteristics is crucial in determining the suitability of the material for a specific application. In this chapter, we will discuss the importance of material characterization in metallurgy.

Importance of Material Characterization

1. Selecting the Right Material for the Right Application:
Material characterization plays a crucial role in selecting the right material for a specific application. By analyzing the mechanical, physical, and chemical properties of different materials, metallurgists can determine which material is best suited for a specific application. For example, the material used in aerospace applications must have high strength, corrosion resistance, and good thermal properties.

2. Quality Control:
Material characterization is essential in ensuring the quality of the material produced. Metallurgists can use characterization techniques to identify the presence of impurities or defects in the material that can affect its properties. This knowledge helps to develop quality control measures that ensure that the produced material meets the required specifications.

3. Understanding the Material Behavior:
Material characterization techniques can provide valuable insights into how the material behaves under specific conditions. For example, the

material's response to stress, temperature, and pressure can be analyzed to determine its suitability for different applications. This understanding of material behavior can help to develop new materials that can perform better in specific applications.

4. Developing and Improving Manufacturing Processes:

Material characterization can also be used to improve and develop new manufacturing processes. By understanding the material's properties, metallurgists can determine the optimal processing conditions that can result in high-quality material with improved properties. This information can also be used to optimize existing manufacturing processes to reduce production costs and increase efficiency.

5. Failure Analysis:

Material characterization is essential in failure analysis investigations. When a material fails in service, material characterization techniques can be used to determine the cause of the failure. This knowledge helps to develop strategies to prevent similar failures in the future.

Common Material Characterization Techniques

Material characterization process involves analyzing the microstructure, surface properties, and chemical composition of the material, among other parameters.

One common technique used in material characterization is microscopy, which allows for the observation of the microstructure of the material. Electron microscopy, in particular, provides high-resolution imaging and can be used to investigate the atomic and electronic structure of the material. Scanning electron microscopy (SEM) and transmission electron microscopy (TEM) are some of the most commonly used electron microscopy techniques.

Another technique used in material characterization is X-ray diffraction (XRD), which helps in determining the crystal structure of the material. This technique works by exposing the material to X-rays and analyzing the pattern of diffracted X-rays. The resulting pattern can be used to identify the crystal structure of the material.

Other techniques used in material characterization include spectroscopy, which is used to analyze the chemical composition of the material, and mechanical testing, which helps in determining the mechanical properties of the material, such as strength and ductility.

Conclusion

In conclusion, material characterization is an essential aspect of metallurgy that plays a crucial role in determining the suitability of materials for specific applications. By understanding the material's properties, metallurgists can develop new materials, improve manufacturing processes, and prevent material failures. The availability of advanced characterization techniques has made it possible to analyze materials more accurately, leading to the development of new and improved materials. The knowledge gained from material characterization will continue to be crucial in the future development of materials for various applications.

Chapter 46 | The Role of Metallurgy in Aerospace Materials Selection

Aerospace engineering requires materials with a unique combination of properties that can withstand high-stress and high-temperature environments. The role of metallurgy in aerospace materials selection is critical as it provides a foundation for identifying and developing materials with the desired properties. Metallurgy plays an essential role in the design, manufacturing, and testing of aerospace materials, ensuring that they meet stringent performance requirements. This chapter will discuss the critical properties required for aerospace materials, the types of metals and alloys used, and the role of metallurgy in their development and selection.

Properties of Aerospace Materials

Aerospace materials must possess several key properties to withstand harsh operating conditions, including high temperatures, high stresses, and extreme weather conditions. The critical properties required for aerospace materials include:

1. High Strength: Aerospace materials must have high strength to withstand the forces exerted on them during flight and in harsh weather conditions.

2. Lightweight: The weight of aircraft and spacecraft is critical as it affects fuel efficiency and payload capacity. Materials with high strength and low weight are, therefore, ideal for aerospace applications.

3. High Temperature Resistance: Aerospace materials must be able to withstand high temperatures generated by friction, atmospheric re-entry, and propulsion systems.

4. Corrosion Resistance: Aerospace materials must be resistant to corrosion as they are exposed to harsh environmental conditions.

5. Fatigue Resistance: Aerospace materials must be able to withstand repeated loading and unloading cycles without failure.

Metals and Alloys Used in Aerospace Applications

Metals and alloys are commonly used in aerospace applications due to their unique properties, including high strength, toughness, and corrosion resistance. The most commonly used metals and alloys in aerospace applications include:

1. Aluminum Alloys: Aluminum alloys are lightweight, corrosion-resistant, and have excellent thermal conductivity, making them ideal for aircraft structures and components.

2. Titanium Alloys: Titanium alloys are known for their high strength-to-weight ratio, excellent corrosion resistance, and ability to withstand high temperatures, making them ideal for aircraft engines and structural components.

3. Nickel-based Superalloys: Nickel-based superalloys have excellent high-temperature resistance and are used in aerospace applications, including turbine engines and exhaust systems.

4. Stainless Steel Alloys: Stainless steel alloys are known for their excellent corrosion resistance and are used in aerospace applications where corrosion resistance is critical.

The Role of Metallurgy in Aerospace Materials Selection

Metallurgy plays a critical role in aerospace materials selection as it ensures that the selected materials possess the required properties to withstand the harsh operating conditions. Metallurgical processes, including alloy development, heat treatment, and surface treatment, are used to enhance the properties of materials, making them suitable for aerospace applications. Metallurgists work closely with engineers to identify materials with the desired properties and optimize their properties through metallurgical processes.

1. **Alloy Development:** Metallurgists play a critical role in alloy development for aerospace applications. They work to develop alloys with the desired properties, including high strength, low weight, and high-temperature resistance, to meet the unique requirements of aerospace applications.

2. **Heat Treatment:** Heat treatment is used to enhance the mechanical and physical properties of materials, including strength, ductility, and toughness. Metallurgists work to optimize heat treatment processes to improve the properties of aerospace materials.

3. **Surface Treatment:** Surface treatment processes, including corrosion-resistant coatings and finishes, are used to protect aerospace materials from corrosion and wear. Metallurgists play a critical role in developing and optimizing surface treatment processes to ensure that materials meet the stringent requirements of aerospace applications.

One example of a critical application in aerospace is jet engine components, which require materials that can withstand high temperatures, stresses, and corrosion. Materials such as nickel-based superalloys, titanium alloys, and ceramic matrix composites (CMCs) have been developed and optimized for use in these components. Metallurgists are involved in the testing and evaluation of these materials, as well as in developing new alloys with improved properties.

Another critical area in aerospace materials selection is the design and manufacture of airframe structures, which must be strong, lightweight, and durable. Aluminum alloys, titanium alloys, and advanced composites are commonly used in these applications. Metallurgists are involved in the characterization and testing of these materials to ensure their suitability for specific applications.

Metallurgy also plays a vital role in the design and development of space exploration vehicles and equipment. Materials used in space must be able to withstand the harsh conditions of space, including extreme temperatures, radiation, and vacuum conditions. Metallurgists are involved in the development of materials for use in spacecraft and rovers, such as lightweight composites, high-strength alloys, and thermal protection systems.

In addition to material selection and development, metallurgists in aerospace are also involved in failure analysis and prevention. The failure of critical components in aerospace applications can have catastrophic consequences, making it essential to understand the root causes of failures and develop strategies to prevent them.

Overall, the role of metallurgy in aerospace materials selection is crucial in ensuring the safe and efficient operation of aircraft, spacecraft, and related equipment. Metallurgists play a vital role in the development, testing, and optimization of materials for use in these applications, as well as in the analysis and prevention of failures. The continued advancement of metallurgical techniques and materials will undoubtedly play an essential role in the future of aerospace engineering and exploration.

Chapter 47 | Best Practices for Reducing Steelmaking Costs

Steel is one of the most commonly used materials in construction, manufacturing, and other industrial applications. However, steel production can be a costly process that requires careful management to maintain profitability. This chapter explores some best practices for reducing steelmaking costs.

1. Use Energy Efficient Technologies

Energy is one of the largest costs associated with steel production. By using energy-efficient technologies, such as electric arc furnaces (EAF) and gas turbines, steelmakers can significantly reduce energy consumption and costs. These technologies not only reduce the use of fossil fuels but also improve the quality of the steel produced.

2. Optimize Raw Material Procurement

Raw materials, such as iron ore, coal, and scrap metal, are essential for steel production. However, the prices of these materials can fluctuate significantly, impacting the overall cost of steelmaking. By optimizing raw material procurement, steelmakers can ensure that they have a consistent supply of raw materials at the best possible prices. This can be achieved by developing long-term relationships with suppliers, diversifying suppliers, and using market intelligence to make informed purchasing decisions.

3. Minimize Scrap

Scrap metal is an important raw material for steel production. However, using too much scrap can lead to increased costs due to the need for additional refining and processing. By minimizing scrap, steelmakers can reduce costs associated with the need for additional processing steps and improve the quality of the final product. This can be achieved by optimizing the use of scrap in the production process and by implementing a scrap tracking and quality control program.

4. Optimize Production Processes

By optimizing production processes, steelmakers can reduce waste, improve product quality, and increase overall efficiency. This can be achieved by implementing lean manufacturing principles, improving equipment maintenance practices, and using advanced process control technologies.

5. Implement a Continuous Improvement Program

Continuous improvement is essential for reducing steelmaking costs. By implementing a continuous improvement program, steelmakers can identify opportunities for cost savings, improve processes, and reduce waste. This can be achieved by using key performance indicators (KPIs) to measure performance, conducting regular audits and assessments, and involving employees in the improvement process.

In conclusion, reducing steelmaking costs is essential for maintaining profitability in the steel industry. By using energy-efficient technologies, optimizing raw material procurement, minimizing scrap, optimizing production processes, and implementing a continuous improvement program, steelmakers can reduce costs and improve profitability.

Chapter 48 | The Importance of Safety in Metallurgical Applications

Metallurgical applications are widespread in various industries, from construction to aerospace, and safety is crucial to the success and well-being of all personnel involved. Safety in metallurgy refers to the measures taken to prevent accidents and protect workers from potential hazards associated with the handling, processing, and transportation of metals and alloys.

Importance of Safety in Metallurgical Applications

The importance of safety in metallurgy cannot be overemphasized. Metallurgical processes involve the use of high temperatures, heavy machinery, and hazardous chemicals, which can pose significant risks to workers. Accidents can result in severe injuries or even death, leading to a loss of skilled workers and reduced productivity. Additionally, workplace accidents can cause damage to equipment and facilities, leading to production delays and increased costs.

Best Practices for Ensuring Safety in Metallurgical Applications

The following best practices can help ensure safety in metallurgical applications:

1. Provide Training: All workers involved in metallurgical processes should receive adequate training on the potential hazards associated with their work, as well as the use of safety equipment and procedures.

2. Use Proper Protective Gear: Protective gear, including helmets, goggles, gloves, and boots, should be provided and used appropriately to protect workers from potential hazards.

3. Maintain Equipment: All equipment used in metallurgical processes should be regularly inspected and maintained to ensure that it is safe and in good working condition.

4. Follow Safety Procedures: Standard operating procedures should be developed and followed to ensure that all tasks are performed safely and efficiently.

5. Implement Hazard Communication: Proper labeling and communication of hazardous materials and equipment can prevent accidents and ensure safe handling.

6. Conduct Regular Safety Audits: Regular safety audits can help identify potential hazards and ensure that all safety procedures are being followed.

7. Encourage a Safety Culture: Encouraging a culture of safety in the workplace, where employees feel empowered to speak up about safety concerns, can help prevent accidents and ensure a safe working environment.

Conclusion

In conclusion, safety in metallurgical applications is critical to the success and well-being of all personnel involved. By implementing best practices for ensuring safety, workers can be protected from potential hazards, and accidents can be prevented. A commitment to safety is essential to maintaining a safe and efficient workplace, improving productivity, and reducing costs associated with accidents and injuries.

Chapter 49 | The Role of Metallurgy in High-Performance Materials Developmen

High-performance materials are at the forefront of technological advancements, enabling the development of cutting-edge products and applications across various industries. Metallurgy, as a fundamental discipline in materials science, plays a crucial role in the design, development, and optimization of high-performance materials. This chapter explores the significant contributions of metallurgy in the development of high-performance materials.

Understanding Material Structure-Property Relationships

Metallurgy provides a deep understanding of the structure-property relationships in materials. Metallurgists study the microstructure of materials at the atomic and microscopic levels, investigating how various elements, processing techniques, and heat treatments influence the material's mechanical, thermal, electrical, and chemical properties. This knowledge is essential for tailoring materials to meet specific performance requirements.

Alloy Design and Optimization

Metallurgists play a key role in alloy design, combining different elements to create materials with enhanced properties. Through alloying, metallurgists can improve strength, corrosion resistance, thermal stability, and other desirable characteristics. They consider factors such as phase diagrams, solubility limits, grain size control, and precipitation hardening to optimize alloy compositions and processing parameters for high-performance materials.

Advanced Processing Techniques

Metallurgy encompasses a wide range of processing techniques that are crucial for high-performance materials development. Metallurgists employ techniques like powder metallurgy, rapid solidification, and severe plastic deformation to produce materials with unique microstructures and enhanced properties. These techniques allow for fine-tuning of material properties, such as improved strength-to-weight ratios, enhanced wear resistance, and superior mechanical performance.

Heat Treatment and Thermal Processing

Heat treatment processes, such as annealing, quenching, and tempering, are essential tools in metallurgy for modifying material properties. Metallurgists utilize their knowledge of phase transformations and diffusion kinetics to design heat treatment cycles that optimize the mechanical, thermal, and chemical properties of high-performance materials. This enables the creation of materials with tailored properties, such as high hardness, improved ductility, or superior thermal stability.

Materials Characterization and Testing

Metallurgists employ a wide range of characterization and testing techniques to assess the performance of high-performance materials. These techniques include microscopy, X-ray diffraction, mechanical testing, and corrosion analysis, among others. By understanding material behavior under different conditions, metallurgists can refine materials and identify any limitations or potential areas for improvement.

Collaboration and Multidisciplinary Approaches

Metallurgy often involves collaboration with other scientific disciplines, such as materials science, chemistry, and engineering. Metallurgists work alongside experts from these fields to develop innovative solutions and integrate materials into high-performance applications. This multidisciplinary approach facilitates the creation of materials with exceptional properties, enabling advancements in fields like aerospace, automotive, electronics, and energy.

Conclusion

In conclusion, metallurgy plays a vital role in the development of high-performance materials. Through understanding material structure-property relationships, alloy design, advanced processing techniques, heat treatment, and materials characterization, metallurgists contribute to the creation of materials with superior properties. The application of metallurgical principles and techniques is instrumental in pushing the boundaries of material performance and enabling technological advancements in various industries.

50 | BEST PRACTICES FOR REFRACTORY MAINTENANCE AND REPAIR

Refractory materials are crucial components in high-temperature industrial processes, providing thermal insulation and protection to equipment and structures. Proper maintenance and repair of refractories are essential to ensure their longevity and optimal performance. This chapter discusses the best practices for refractory maintenance and repair to enhance their durability and reliability.

1. Regular Inspection

Regular inspections are critical to identify signs of wear, erosion, cracking, or damage in refractory linings. Inspections should be carried out at scheduled intervals and after each campaign or shutdown. Visual inspections, thermal imaging, and non-destructive testing techniques can help detect issues at an early stage and prevent potential failures.

2. Cleaning and Preparations

Before performing any maintenance or repair work on refractories, it is crucial to clean the surfaces thoroughly. This includes removing any debris, loose materials, or contaminants. Proper cleaning ensures better adhesion of repair materials and improves the effectiveness of maintenance procedures.

3. Selection of Repair Materials

Choosing the right repair materials is essential for effective refractory maintenance. The selection should be based on factors such as the operating temperature, chemical environment, abrasion resistance, and thermal expansion properties. Using high-quality repair materials that are compatible with the original refractory will ensure optimal performance and prolong the refractory's service life.

4. Correct Application Techniques

Applying repair materials correctly is crucial for achieving long-lasting repairs. It is essential to follow manufacturer guidelines and recommended application techniques. This includes proper mixing of repair materials, achieving the desired consistency, and applying the materials evenly and uniformly. The use of appropriate tools and equipment will help ensure accurate and efficient repairs.

5. Thermal Curing

Many repair materials require thermal curing to achieve their desired strength and properties. It is important to follow the recommended curing procedures, including heating and cooling rates, to prevent cracking or damage due to thermal stresses. Properly controlled thermal curing enhances the bond between the repair material and the existing refractory lining.

6. Post-Repair Inspection

After completing repairs, conducting post-repair inspections is necessary to verify the effectiveness of the repairs. Inspections may include visual examination, non-destructive testing, or performance testing under operating conditions. This step ensures that the repaired refractories meet the desired performance standards and are ready for service.

7. Preventive Maintenance

Implementing a preventive maintenance program is crucial for prolonging the lifespan of refractories. Regular maintenance activities such as cleaning, inspections, and minor repairs should be performed as per the maintenance schedule. This proactive approach helps identify and address potential issues before they escalate, ensuring the optimal performance and longevity of refractories.

Conclusion

Maintaining and repairing refractories using best practices is essential for ensuring their longevity, performance, and safety in high-temperature applications. Regular inspections, proper cleaning, selection of appropriate repair materials, correct application techniques, thermal curing, post-repair inspections, and preventive maintenance are key elements of effective refractory maintenance and repair. By following these best practices, industries can minimize downtime, extend refractory service life, and optimize the efficiency of their operations.

Chapter 51 | Emerging Trends in Additive Manufacturing of Metals

Additive manufacturing, also known as 3D printing, has revolutionized the manufacturing industry, including the production of metal components. This chapter explores the emerging trends in additive manufacturing of metals, highlighting the advancements, challenges, and potential applications in various industries.

1. Multi-Material Printing

One of the emerging trends in metal additive manufacturing is the ability to print components with multiple materials. This allows for the creation of complex, functional parts with tailored properties. Multi-material printing enables the incorporation of different metals, alloys, or even non-metallic materials within a single component, opening up new possibilities in design and functionality.

2. High-Speed Printing

Advancements in printing technologies and process optimization have led to high-speed metal additive manufacturing. Faster printing speeds significantly reduce production time and enhance the efficiency of manufacturing operations. High-speed printing enables rapid prototyping, on-demand production, and scalability in various industries, including aerospace, automotive, and medical.

3. Improved Material Selection

As the field of metal additive manufacturing matures, there is a growing emphasis on expanding the range of printable materials. Research and development efforts focus on developing new metal alloys specifically designed for additive manufacturing, offering enhanced mechanical properties, corrosion resistance, and heat resistance. The ability to print a wider range of metals expands the potential applications and end-use industries.

4. In-Situ Monitoring and Quality Control

Ensuring the quality and integrity of printed metal components is crucial. Emerging trends in metal additive manufacturing involve the integration of in-situ monitoring systems to detect and address defects or inconsistencies during the printing process. Real-time monitoring techniques, such as thermal imaging, sensors, and spectroscopy, enable continuous quality control and optimization of printing parameters.

5. Post-Processing Techniques

Post-processing plays a vital role in refining and improving the properties of printed metal components. Emerging trends focus on developing innovative post-processing techniques, such as heat treatments, surface finishing, and machining, specifically tailored for additive manufacturing. These techniques enhance the mechanical properties, surface quality, and dimensional accuracy of printed metal parts.

6. Design Optimization and Simulation

Advancements in design software and simulation tools are empowering engineers to optimize designs for additive manufacturing. The use of generative design algorithms and topology optimization techniques allows for lightweight, high-performance structures that can only be produced through additive manufacturing. Simulation tools also aid in predicting the behavior and performance of printed metal components before fabrication.

7. Integration with Traditional Manufacturing:

An emerging trend in metal additive manufacturing is the integration of additive processes with traditional manufacturing techniques. Hybrid manufacturing approaches combine the strengths of additive manufacturing, such as design flexibility and complexity, with the efficiency and cost-effectiveness of traditional processes like machining and forging. This integration enables the production of complex, functional metal parts with enhanced efficiency and reduced material waste.

Conclusion

Emerging trends in additive manufacturing of metals are transforming the way we design, produce, and utilize metal components. Multi-material printing, high-speed printing, improved material selection, in-situ monitoring, post-processing techniques, design optimization, and integration with traditional manufacturing are driving advancements in this field. These trends hold immense potential for various industries, enabling the production of customized, lightweight, and high-performance metal parts. As research and development efforts continue, the future of metal additive manufacturing looks promising, with widespread adoption and new applications on the horizon.

CHAPTER 52 | SUSTAINABLE METALLURGY: CHALLENGES AND OPPORTUNITIES

Sustainable metallurgy is a growing field that addresses the environmental, social, and economic impacts of metallurgical processes. This chapter explores the challenges and opportunities associated with achieving sustainability in the metallurgical industry. It highlights the importance of adopting sustainable practices and technologies to minimize resource depletion, reduce environmental footprint, and create a positive societal impact.

1. Resource Efficiency

One of the key challenges in sustainable metallurgy is optimizing resource utilization. This involves minimizing raw material consumption, reducing energy requirements, and increasing material recovery and recycling rates. Advanced process technologies, such as process integration, waste heat recovery, and material recycling, play a crucial role in improving resource efficiency and reducing the environmental impact of metallurgical operations.

2. Carbon Footprint Reduction

Reducing greenhouse gas emissions is a significant challenge in sustainable metallurgy. The industry is exploring various strategies, such as energy-efficient process design, use of renewable energy sources, carbon capture and storage, and carbon offset programs. Implementing these measures not only helps mitigate climate change but also improves the industry's social and environmental performance.

3. Water Management

Water scarcity and pollution are pressing concerns in many regions.

Sustainable metallurgy aims to minimize water consumption, optimize water reuse and recycling, and implement effective wastewater treatment technologies. Water management strategies, such as closed-loop systems, water-saving process designs, and responsible water sourcing, are crucial for sustainable metallurgical operations.

4. Circular Economy

Transitioning to a circular economy model is a significant opportunity for sustainable metallurgy. This involves designing products and processes with the intention of maximizing material recovery and reuse. Implementing closed-loop material flows, product life extension, and remanufacturing practices contribute to resource conservation and waste reduction, thereby promoting sustainability in the metallurgical industry.

5. Responsible Sourcing and Supply Chain

Sustainable metallurgy involves considering the entire supply chain, from raw material extraction to end-of-life product management. Ensuring responsible sourcing practices, ethical mineral extraction, and fair labor conditions are essential for sustainable metallurgical operations. Collaboration with suppliers, certification programs, and traceability initiatives play a vital role in promoting responsible sourcing in the industry.

6. Stakeholder Engagement and Social Impact

Sustainable metallurgy recognizes the importance of engaging with stakeholders, including local communities, employees, and regulatory bodies. Establishing transparent communication channels, addressing social concerns, and incorporating local knowledge and expertise contribute to the positive social impact of metallurgical operations. Engaging stakeholders also helps build trust and foster collaboration in sustainable decision-making processes.

7. Innovation and Research

Innovation and research are key drivers of sustainable metallurgy.

Encouraging research and development efforts in clean technologies, alternative processes, and eco-friendly materials opens up opportunities for more sustainable metallurgical practices. Collaboration between academia, industry, and governmental organizations promotes knowledge exchange and drives the development of sustainable solutions.

Conclusion

Achieving sustainability in metallurgy is a multifaceted challenge that requires concerted efforts from industry, academia, and policymakers. By addressing challenges such as resource efficiency, carbon footprint reduction, water management, circular economy, responsible sourcing, stakeholder engagement, and fostering innovation, the metallurgical industry can seize opportunities to transition towards more sustainable practices. Embracing sustainable metallurgy not only reduces environmental impacts but also enhances social and economic benefits, creating a more resilient and responsible industry for the future.

CHAPTER 53 | NOVEL APPROACHES TO EXTRACTING METALS FROM ORES

Extracting metals from ores is a fundamental process in metallurgy, and advancements in extraction technologies can have significant implications for the industry. This chapter explores novel approaches to extracting metals from ores, highlighting innovative techniques and processes that offer potential improvements in efficiency, sustainability, and resource utilization.

1. Bioleaching

Bioleaching is a promising approach that utilizes microorganisms to extract metals from ores. Certain bacteria and fungi have the ability to catalyze the dissolution of metals from ores, making them more accessible for recovery. Bioleaching offers advantages such as low energy requirements, reduced environmental impact, and the potential to extract metals from low-grade or complex ores.

2. Hydrometallurgical Processes

Hydrometallurgical processes involve the use of aqueous solutions to extract metals from ores. Novel approaches in hydrometallurgy include innovative leaching agents, such as non-toxic or environmentally friendly solvents, as well as the development of advanced separation and purification techniques. These advancements aim to improve the efficiency of metal extraction while minimizing the use of hazardous chemicals.

3. Electrochemical Extraction

Electrochemical extraction techniques utilize electrochemical reactions to selectively extract metals from ores. These approaches involve the use of specific electrodes and controlled electrochemical potentials to facilitate metal dissolution and deposition. Electrochemical extraction methods offer advantages such as higher selectivity, reduced energy consumption, and the ability to recover metals in high purity.

4. Microwave-Assisted Extraction

Microwave-assisted extraction is an emerging technique that utilizes microwave energy to enhance the extraction of metals from ores. The application of microwaves can generate localized heating, promoting faster and more efficient dissolution of metals. This approach offers potential benefits in terms of reduced energy consumption, shorter processing times, and improved metal recovery rates.

5. Ionic Liquids

Ionic liquids are non-volatile, molten salts that have unique properties, including high solvating capabilities. They are being explored as alternative solvents for metal extraction from ores. Ionic liquids offer advantages such as lower toxicity, improved selectivity, and the ability to dissolve a wide range of metal species. Their tunable properties make them suitable for extracting specific metals from complex ore matrices.

6. Solvent Extraction and Ion Exchange

Solvent extraction and ion exchange processes are widely used in metal extraction, but novel approaches are being developed to enhance their efficiency and selectivity. Advances in extractant design, optimization of operating conditions, and the use of alternative organic and inorganic phases contribute to improved metal extraction and separation processes.

7. Nanostructured Materials

The utilization of nanostructured materials in metal extraction processes shows great promise. Nanomaterials can provide large surface areas, enhanced reactivity, and tailored properties, making them suitable for efficient metal extraction from ores. Novel approaches include the use of nanoparticles, nanocomposites, and nanostructured membranes for enhanced leaching, separation, and recovery of metals.

Conclusion

Novel approaches to extracting metals from ores hold significant potential for improving the efficiency, sustainability, and selectivity of metal extraction processes. Bioleaching, hydrometallurgical processes, electrochemical extraction, microwave-assisted extraction, ionic liquids, solvent extraction, ion exchange, and the utilization of nanostructured materials are just a few examples of the innovative techniques being explored. By embracing these novel approaches, the metallurgical industry can enhance resource utilization, reduce environmental impacts, and optimize the extraction of valuable metals from ores. Continued research, development, and collaboration in this field will drive further advancements and pave the way for a more efficient and sustainable metal extraction industry.

CHAPTER 54 | THE ADVANCEMENTS AND LIMITATIONS OF NON-DESTRUCTIVE TESTING TECHNIQUES IN METALLURGY

Non-destructive testing (NDT) techniques play a crucial role in the metallurgical industry for assessing the quality, integrity, and properties of materials and components without causing damage. This chapter explores the advancements and limitations of various NDT techniques used in metallurgy, highlighting their benefits, challenges, and areas of application.

1. Ultrasonic Testing (UT)

UT utilizes high-frequency sound waves to detect and characterize internal flaws, such as cracks, voids, and inclusions, in metallic materials. Advancements in UT include phased array ultrasonics and time-of-flight diffraction, enabling improved flaw detection, sizing, and imaging. However, limitations include the dependence on material properties, limited penetration in highly attenuative materials, and the need for skilled operators for accurate interpretation of results.

2. Radiographic Testing (RT)

RT involves the use of X-rays or gamma rays to inspect the internal structure of materials for defects, such as cracks, porosity, and inclusions. Digital radiography and computed tomography have advanced RT capabilities, providing enhanced imaging resolution, data analysis, and 3D visualization. Limitations of RT include the need for radiation safety measures, limited sensitivity to certain types of defects, and the challenge of interpreting complex images.

3. Magnetic Particle Testing (MT)

MT detects surface and near-surface defects in ferromagnetic materials by applying a magnetic field and observing the response of magnetic particles. Advances in MT include the use of portable and automated systems, fluorescent particles for improved visibility, and remote field testing for inspecting thick-walled components. Limitations of MT include its applicability only to ferromagnetic materials and the need for proper surface preparation to ensure reliable results.

4. Liquid Penetrant Testing (PT)

PT is used to identify surface-breaking defects in materials by applying a liquid penetrant and observing its capillary action into the defects. Advances in PT include the development of highly sensitive fluorescent dyes, visible dye penetrants, and portable inspection systems. However, PT has limitations in detecting subsurface defects and is restricted to accessible surfaces.

5. Eddy Current Testing (ECT):

ECT uses electromagnetic induction to detect and characterize surface and near-surface defects, such as cracks and corrosion, in conductive materials. Advancements in ECT include the use of multi-frequency and array probes, as well as imaging techniques for enhanced defect detection and discrimination. Limitations of ECT include the dependency on electrical conductivity, sensitivity to probe lift-off, and limited penetration depth.

6. Acoustic Emission Testing (AET)

AET detects and analyzes acoustic signals generated by the rapid release of energy within materials. It is effective in monitoring crack growth and detecting defects under dynamic loading conditions. Advances in AET include the use of sensor arrays, signal processing algorithms, and pattern recognition techniques for accurate defect identification. However, AET has limitations in localizing defect positions precisely and differentiating between various types of signals.

Advancements in NDT Technology

Advancements in NDT technology include the integration of artificial intelligence, machine learning, and data analytics for automated defect recognition, real-time monitoring, and predictive maintenance. Digitalization and remote inspection capabilities also enhance the efficiency and accessibility of NDT processes.

Limitations of NDT techniques in metallurgy include:

1. Interpretation Challenges: NDT results often require expert interpretation to accurately identify and characterize defects. This requires skilled personnel who are knowledgeable about the specific technique and the materials being inspected.

2. Surface Accessibility: Some NDT techniques are limited to inspecting accessible surfaces, making it challenging to assess the integrity of internal or hidden components. Specialized techniques or supplementary methods may be needed for comprehensive inspection.

3. Material Dependency: Certain NDT techniques are more suitable for specific materials. For example, magnetic particle testing is effective only for ferromagnetic materials, while eddy current testing is limited to conductive materials. Material properties, such as electrical conductivity and permeability, can affect the reliability and sensitivity of the NDT results.

4. Limited Penetration Depth: The ability of NDT techniques to detect defects is influenced by the penetration depth of the applied energy or probe. For example, ultrasonic waves have limited penetration in highly attenuative materials, limiting their effectiveness for thick or highly absorbent components.

5. Detection Sensitivity: NDT techniques may have limitations in detecting certain types of defects, particularly small or subsurface flaws. Factors such as the size, orientation, and location of defects can affect the detectability, requiring complementary techniques or multiple inspections for comprehensive evaluation.

6. Environmental Constraints: Some NDT techniques require controlled environmental conditions, such as temperature, humidity, or cleanliness, for accurate results. Adverse environmental factors may hinder the effectiveness or reliability of the inspections.

7. Equipment and Training Requirements: NDT techniques often require specialized equipment, which can be costly to acquire, maintain, and calibrate. Furthermore, proper training and certification of personnel are necessary to ensure accurate and consistent results.

8. Time and Cost Considerations: NDT inspections can be time-consuming, especially for large structures or complex components. The need for skilled operators, equipment setup, and data analysis may result in increased costs and longer inspection durations.

9. Limitations in Quantitative Assessment: While NDT techniques provide valuable qualitative information about defects, they may have limitations in providing precise quantitative data, such as defect size, depth, or severity. Supplementary techniques or complementary methods may be required for accurate quantification.

Conclusion

Non-destructive testing techniques continue to advance, enabling better assessment of material quality and component integrity in the metallurgical industry. Ultrasonic testing, radiographic testing, magnetic particle testing, liquid penetrant testing, eddy current testing, acoustic emission testing, and emerging technologies offer valuable insights into the condition of materials without causing damage. While these techniques have made significant progress, it is essential to acknowledge their limitations and ensure proper application and interpretation of results. By leveraging the advancements and addressing the

Chapter 55 | Innovative Coating Technologies for Corrosion Protection in Metallurgical Applications

Corrosion is a significant concern in metallurgical applications, leading to material degradation, structural failures, and economic losses. This chapter explores innovative coating technologies that provide effective corrosion protection in various metallurgical industries. These advanced coatings offer enhanced durability, extended service life, and improved performance compared to traditional corrosion protection methods.

1. Ceramic Coatings

Ceramic coatings, such as thermal barrier coatings and ceramic-metallic hybrid coatings, offer excellent resistance to corrosion, high temperatures, and wear. They are widely used in industries like aerospace, automotive, and power generation. Advanced techniques like plasma spraying, chemical vapor deposition, and sol-gel methods allow precise control over coating properties, thickness, and composition, resulting in superior corrosion protection.

2. Polymer Coatings

Polymer coatings provide a versatile and cost-effective solution for corrosion protection. Advanced polymer coatings, including epoxy, polyurethane, and fluoropolymer coatings, offer excellent chemical resistance, adhesion, and durability. The development of nanocomposite coatings incorporating nanoparticles and smart polymers has further enhanced their corrosion protection capabilities.

3. Conversion Coatings

Conversion coatings, such as chromate, phosphate, and oxide coatings, chemically convert the metal surface to a more corrosion-resistant form. They are commonly used in the aerospace, automotive, and electronics industries. Advances in conversion coating processes, such as environmentally friendly alternatives to hexavalent chromate, have improved both the performance and sustainability of these coatings.

4. Thin Film Coatings

Thin film coatings, including physical vapor deposition (PVD) and chemical vapor deposition (CVD) coatings, offer exceptional corrosion resistance, hardness, and surface smoothness. These coatings are widely employed in cutting tools, electronic components, and automotive applications. Innovative approaches, such as atomic layer deposition (ALD) and magnetron sputtering, allow precise control over coating thickness, composition, and microstructure.

5. Self-Healing Coatings

Self-healing coatings utilize encapsulated corrosion inhibitors or microcapsules containing healing agents to autonomously repair coating damage and prevent corrosion propagation. These coatings are particularly valuable in challenging environments and critical applications. Self-healing mechanisms, such as corrosion-induced or temperature-induced healing, provide long-term protection and reduce maintenance requirements.

6. Smart Coatings

Smart coatings integrate functionalities such as corrosion sensing, self-cleaning, or stimuli-responsive properties to enhance corrosion protection. These coatings can detect early signs of corrosion, release corrosion inhibitors, or repel contaminants to maintain the integrity of the coated surface. The incorporation of sensors, nanomaterials, and responsive polymers enables real-time monitoring and active corrosion prevention.

7. Bio-inspired Coatings

Bio-inspired coatings draw inspiration from nature to develop innovative corrosion protection strategies. For example, biomimetic coatings mimic the self-cleaning properties of lotus leaves or the protective mechanisms of certain marine organisms. These coatings offer enhanced resistance to fouling, corrosion, and environmental degradation.

8. Hybrid and Multilayer Coatings

Hybrid and multilayer coatings combine different materials, such as metals, ceramics, and polymers, to synergistically enhance corrosion protection. These coatings leverage the unique properties of each layer to provide superior performance in terms of corrosion resistance, mechanical strength, and adhesion. Advanced design strategies and characterization techniques enable tailored coating architectures for specific applications.

Conclusion

Innovative coating technologies play a vital role in mitigating corrosion-related challenges in metallurgical applications. Ceramic coatings, polymer coatings, conversion coatings, thin film coatings, self-healing coatings, smart coatings, bio-inspired coatings, and hybrid/multilayer coatings offer a diverse range of options for effective corrosion protection. Continued research and development efforts in coating materials, manufacturing processes, and performance

Chapter 56 | The Evolution of Metallurgy: From Ancient Times to Modern Industry

Metallurgy, the study and practice of extracting, refining, and manipulating metals, has played a vital role in human civilization since ancient times. This chapter explores the rich history and evolution of metallurgy, from its humble beginnings to its crucial role in modern industry.

1. Early Beginnings

The origins of metallurgy can be traced back to prehistoric times when humans first discovered and utilized native metals like gold and copper. This section explores the early techniques used to extract and shape these metals, including simple forging and casting methods.

2. Bronze Age and the Rise of Alloying

The Bronze Age marked a significant advancement in metallurgy with the discovery of copper and tin alloying, leading to the production of bronze. This section delves into the technological advancements, societal impact, and trade routes associated with the use of bronze in ancient civilizations.

3. Iron Age and Iron Smelting

The Iron Age revolutionized metallurgy with the widespread adoption of iron as a metal. This section examines the development of iron smelting techniques, including the transition from bloomery furnaces to more efficient blast furnaces, and the transformative impact of iron on warfare, agriculture, and construction.

4. Medieval Metallurgy and Alchemy

During the medieval period, metallurgical knowledge expanded, driven by advancements in alchemy and the search for the Philosopher's Stone. This section explores the emergence of new techniques, such as the water-powered hammer, the introduction of steel production, and the rise of guilds and metallurgical centers.

5. Industrial Revolution and Modern Metallurgy

The Industrial Revolution marked a turning point in metallurgy, fueled by advancements in science, engineering, and manufacturing processes. This section discusses the development of crucial techniques like Bessemer and open-hearth steelmaking, the discovery of new alloys, and the impact of metallurgy on the expansion of industries such as railways, construction, and machinery.

6. Materials Science and Technological Innovations

In the modern era, metallurgy evolved into the interdisciplinary field of materials science, integrating physics, chemistry, and engineering principles. This section highlights the impact of scientific discoveries on metallurgy, including the understanding of crystal structures, phase transformations, and the development of advanced alloy design and processing techniques.

7. Metallurgy in the Digital Age

The digital age has brought further advancements to metallurgy, with the integration of computer simulations, data analytics, and automation in materials research and manufacturing processes. This section explores the use of computational modeling, additive manufacturing, and advanced characterization techniques in pushing the boundaries of metallurgical science and engineering.

8. Sustainable Metallurgy

The chapter concludes by discussing the current focus on sustainable metallurgy, highlighting initiatives to minimize the environmental impact of metallurgical processes, promote recycling and circular economy principles, and develop new materials with improved performance and reduced carbon footprint.

Conclusion

The evolution of metallurgy has been a remarkable journey, from ancient civilizations smelting metals by the fire to modern industries employing cutting-edge technologies. Understanding the historical development of metallurgy provides valuable insights into the progression of human knowledge and the profound impact of metallurgical advancements on society, technology, and industrial progress.

Chapter 57 | The Future of Steel: New Materials, Processes, and Applications

Steel, a versatile and essential material in various industries, continues to evolve and adapt to meet the changing demands of the modern world. This chapter explores the future of steel, focusing on the development of new materials, innovative manufacturing processes, and emerging applications that will shape its trajectory in the coming years.

Advanced Steel Alloys

The future of steel lies in the development of advanced alloys with enhanced properties. This section discusses the exploration of new alloying elements, microstructural design, and the use of computational modeling and high-throughput experimentation to create steels with improved strength, corrosion resistance, and performance at high temperatures.

Nanostructured and Amorphous Steels

The manipulation of steel at the nanoscale opens up new possibilities for improved mechanical and functional properties. This section explores the use of nanostructuring techniques, such as severe plastic deformation and nano-additives, to enhance strength, ductility, and wear resistance. Additionally, the development of amorphous steels with unique properties and processing challenges is also discussed.

Sustainable Steel Production

As sustainability becomes a top priority, the steel industry is striving to reduce its environmental footprint. This section explores the development of cleaner production processes, such as hydrogen-based steelmaking and carbon capture and utilization, to minimize greenhouse gas emissions. Additionally, the incorporation of recycled materials and circular economy principles in steel production is also addressed.

Additive Manufacturing of Steel

Additive manufacturing, or 3D printing, is revolutionizing the production of complex components. This section examines the advancements in additive manufacturing techniques specific to steel, including laser powder bed fusion and directed energy deposition. The benefits, challenges, and potential applications of 3D-printed steel components in various industries are discussed.

Smart and Functional Steels

The future of steel goes beyond traditional structural applications. This section explores the development of smart and functional steels that possess unique properties, such as shape memory, self-healing, or self-cleaning capabilities. These steels have the potential to revolutionize industries like automotive, aerospace, and energy storage by providing innovative solutions to complex engineering challenges.

Steel in Sustainable Infrastructure

Steel plays a vital role in building sustainable infrastructure. This section discusses the use of advanced steels in construction, transportation, and renewable energy applications. Topics include high-strength and lightweight steels for efficient structures, corrosion-resistant steels for marine environments, and steel composites for seismic-resistant buildings.

Steel in Biomedical and Biotechnological Applications

The biomedical field offers exciting opportunities for steel. This section explores the use of biocompatible and bioresorbable steels in medical implants, surgical instruments, and drug delivery systems. The challenges associated with corrosion resistance, wear properties, and biocompatibility are addressed, along with the potential of steel in tissue engineering and regenerative medicine.

Steel in Energy Storage and Conversion

The demand for energy storage and conversion technologies is growing rapidly. This section discusses the potential of steel in applications such as batteries, fuel cells, and hydrogen storage. The development of steel-based materials with improved conductivity, catalytic activity, and durability is explored, along with the challenges of integrating steel into emerging energy technologies.

Conclusion

The future of steel is full of promise, driven by advancements in materials science, manufacturing processes, and emerging applications. The development of advanced alloys, nanostructured steels, sustainable production methods, and innovative applications will shape the steel industry in the coming years. By embracing these advancements, steel will continue to be a fundamental material in addressing the evolving needs of society, from sustainable infrastructure to advanced healthcare and clean energy solutions.

CHAPTER 58 | COMPUTATIONAL MODELING AND SIMULATION IN METALLURGICAL RESEARCH

Computational modeling and simulation have become indispensable tools in metallurgical research, enabling scientists and engineers to understand, predict, and optimize complex metallurgical processes. This chapter explores the use of computational modeling techniques in various aspects of metallurgical research, highlighting their benefits, challenges, and future prospects.

1. Fundamentals of Computational Modeling:

This section provides an overview of computational modeling principles, including the use of mathematical equations, numerical methods, and computer algorithms to simulate metallurgical phenomena. It discusses the advantages of computational modeling in terms of cost-effectiveness, time efficiency, and the ability to explore scenarios that may be challenging or impractical in experimental studies.

2. Thermodynamic Modeling and Phase Equilibrium:

Computational thermodynamics plays a crucial role in predicting phase equilibria, alloy behavior, and phase transformations in metallurgical systems. This section explores the application of thermodynamic databases, such as CALPHAD (Calculation of Phase Diagrams), and the use of thermodynamic modeling software to simulate and analyze complex phase diagrams and phase stability.

3. Kinetic Modeling and Phase Transformations:

Understanding the kinetics of phase transformations is essential in metallurgy. This section discusses the use of kinetic modeling techniques, including diffusion equations, nucleation and growth models, and phase field methods, to predict and simulate the evolution of microstructures during phase transformations, such as solidification, precipitation, and recrystallization.

4. Computational Fluid Dynamics (CFD) in Metallurgical Processes:

Computational fluid dynamics has revolutionized the understanding and optimization of fluid flow and heat transfer in metallurgical processes. This section explores the application of CFD techniques to simulate and analyze complex phenomena, such as flow patterns in ladles and tundishes, gas bubble behavior in steelmaking, and heat transfer in continuous casting.

5. Finite Element Method (FEM) in Structural Analysis:

The finite element method is widely used in structural analysis and mechanical behavior prediction of metallurgical components. This section discusses the application of FEM in simulating mechanical deformation, stress distribution, and failure analysis of materials and structures, including rolling processes, forging operations, and component design optimization.

6. Multi-Scale Modeling Approaches:

Metallurgical systems often exhibit multi-scale phenomena, from atomistic interactions to macroscopic behavior. This section explores multi-scale modeling approaches, such as molecular dynamics, Monte Carlo simulations, and concurrent multi-scale methods, which bridge the gap between different length and time scales to provide a comprehensive understanding of metallurgical processes.

7. Data-Driven Modeling and Machine Learning:

The integration of data-driven modeling and machine learning techniques has revolutionized metallurgical research. This section discusses the use of data analytics, pattern recognition, and machine learning algorithms to extract insights from large datasets, predict material properties, optimize processing parameters, and guide materials discovery and design.

Challenges and Future Perspectives:

The chapter concludes by addressing the challenges and future directions of computational modeling in metallurgical research. It discusses the need for accurate input data, validation against experimental results, computational efficiency, and the integration of models across different length and time scales. It also explores emerging trends, such as the coupling of experimental and computational techniques and the incorporation of uncertainty quantification in modeling approaches.

Conclusion

Computational modeling and simulation have transformed the field of metallurgical research, enabling scientists and engineers to gain valuable insights, optimize processes, and accelerate materials development. From thermodynamic predictions to fluid flow simulations and multi-scale modeling, the integration of computational techniques offers unprecedented opportunities to advance metallurgy. By continuing to refine and expand these modeling approaches, the metallurgical community will unlock new possibilities for innovation, efficiency, and sustainability in the field.

Chapter 59 | The Impact of Environmental Regulations on Metallurgical Industry Practices

In recent years, environmental regulations have significantly influenced the practices and operations of the metallurgical industry. This chapter explores the evolving landscape of environmental regulations and their impact on various aspects of metallurgical processes, highlighting the challenges, opportunities, and strategies employed by the industry to ensure compliance and promote sustainable practices.

Overview of Environmental Regulations

This section provides an overview of the key environmental regulations that affect the metallurgical industry, including air quality standards, water pollution controls, waste management requirements, and greenhouse gas emissions regulations. It discusses the objectives and goals of these regulations in mitigating environmental impacts and protecting human health.

Air Emissions Control

Metallurgical processes can generate significant air pollutants, such as particulate matter, sulfur dioxide, nitrogen oxides, and volatile organic compounds. This section examines the measures employed by the industry to control and reduce air emissions, including the use of emission control technologies, process optimization, and alternative energy sources.

Water Management and Treatment

Water is a crucial resource in metallurgical operations, and its management and treatment are vital for minimizing environmental impacts. This section discusses the implementation of water conservation strategies, wastewater treatment technologies, and the adoption of closed-loop systems to reduce water consumption and protect water quality.

Waste Management and Recycling

Metallurgical processes generate various types of waste, including slag, dust, and spent refractories. This section explores the industry's efforts to minimize waste generation, promote recycling and reuse, and adopt sustainable waste management practices. It also highlights the advancements in waste treatment technologies, such as pyrometallurgical and hydrometallurgical processes.

Energy Efficiency and Carbon Footprint Reduction

Energy consumption is a significant contributor to environmental impacts in the metallurgical industry. This section examines the measures taken to improve energy efficiency, such as process optimization, waste heat recovery, and the use of renewable energy sources. It also explores the industry's efforts to reduce its carbon footprint through emission reduction initiatives and carbon capture and storage technologies.

Life Cycle Assessment and Sustainability

Assessing the environmental impacts of metallurgical processes throughout their life cycle is essential for promoting sustainability. This section discusses the application of life cycle assessment methodologies to evaluate the environmental footprint of metallurgical products and processes. It also explores sustainable design principles, eco-labeling, and the concept of circular economy in the metallurgical industry.

Compliance and Regulatory Challenges

Complying with environmental regulations presents challenges for the metallurgical industry, including technological limitations, economic considerations, and the need for continuous improvement. This section explores the strategies employed by companies to navigate these challenges, such as collaboration with regulatory agencies, investment in research and development, and the adoption of best practices.

Innovation and Future Perspectives

Environmental regulations drive innovation in the metallurgical industry, fostering the development of cleaner technologies and sustainable practices. This section highlights the emerging trends and technologies, such as green metallurgy, process intensification, and the use of advanced control systems and data analytics to optimize environmental performance. It also discusses the importance of industry collaboration, knowledge sharing, and continuous improvement in meeting future regulatory requirements.

Conclusion

Environmental regulations have significantly shaped the practices and operations of the metallurgical industry, driving the adoption of sustainable and environmentally responsible approaches. By embracing technological advancements, implementing efficient waste management strategies, and optimizing energy consumption, the industry can not only comply with regulations but also contribute to a cleaner and more sustainable future. With ongoing innovation and collaboration, the metallurgical industry can continue to thrive while minimizing its environmental footprint and ensuring the well-being of communities and ecosystems.

CHAPTER 60 | IMPROVING THE SUSTAINABILITY OF METALLURGICAL PROCESSES: CHALLENGES AND OPPORTUNITIES

The pursuit of sustainability has become increasingly important in the field of metallurgy. This chapter explores the challenges and opportunities associated with improving the sustainability of metallurgical processes. It delves into the environmental, social, and economic aspects of sustainability and discusses strategies and innovations that can lead to more sustainable practices in the metallurgical industry.

Understanding Sustainability in Metallurgy

This section provides an overview of sustainability principles and their application to metallurgical processes. It explores the triple bottom line approach, which considers environmental, social, and economic factors, and emphasizes the importance of balancing these aspects for long-term sustainability.

Environmental Challenges and Innovations

Metallurgical processes often have significant environmental impacts, such as resource depletion, energy consumption, and emissions of pollutants. This section discusses the challenges faced by the industry in reducing these impacts and explores innovative approaches, including cleaner production techniques, energy-efficient technologies, and the use of renewable resources.

Social Considerations and Community Engagement

Sustainability in metallurgy extends beyond environmental concerns to encompass social aspects such as worker safety, community engagement, and responsible sourcing. This section examines the challenges of ensuring worker health and safety, fostering positive relationships with local communities, and implementing ethical sourcing practices.

Circular Economy and Material Efficiency

The concept of the circular economy is gaining traction in the metallurgical industry, aiming to maximize resource utilization and minimize waste generation. This section explores strategies for improving material efficiency, such as recycling and reusing metal scraps, optimizing material selection, and designing products for recyclability.

Life Cycle Assessment and Environmental Footprinting

Assessing the environmental impacts of metallurgical processes throughout their life cycle is crucial for identifying areas of improvement. This section discusses the application of life cycle assessment methodologies and environmental footprinting techniques to evaluate the sustainability performance of metallurgical processes and products.

Supply Chain Sustainability

Metallurgical processes rely on complex supply chains that span multiple regions and involve various stakeholders. This section explores the challenges and opportunities for improving supply chain sustainability, including responsible sourcing of raw materials, reducing transportation emissions, and promoting transparency and traceability.

Collaboration and Industry Initiatives

Addressing sustainability challenges in metallurgy requires collaboration among industry stakeholders, research institutions, and regulatory bodies. This section discusses the importance of industry initiatives, such as sustainability certifications, industry associations, and collaborative research projects, in driving sustainable practices and fostering knowledge sharing.

Economic Viability and Return on Investment

Sustainability efforts in metallurgy must also consider economic viability and the potential return on investment. This section explores the economic challenges associated with implementing sustainable practices and highlights the business opportunities and cost-saving potential that can arise from improved resource efficiency, waste reduction, and innovation.

Policy and Regulatory Frameworks

Effective policy and regulatory frameworks play a vital role in promoting sustainable practices in the metallurgical industry. This section examines the importance of supportive policies, incentives for innovation, and harmonized regulations that encourage the adoption of sustainable technologies and practices.

Future Perspectives and Emerging Technologies

The journey towards improved sustainability in metallurgical processes is an ongoing endeavor. This section discusses future perspectives and emerging technologies that hold promise for further enhancing sustainability, such as the use of renewable energy sources, advanced process control systems, and breakthrough innovations in materials science.

Conclusion

Improving the sustainability of metallurgical processes is a multifaceted challenge that requires a holistic approach encompassing environmental, social, and economic considerations. By embracing innovation, collaboration, and responsible practices, the metallurgical industry can contribute to a more sustainable future. Overcoming the challenges and seizing the opportunities in sustainability will not only enhance the industry's environmental performance but also foster resilience, competitiveness, and societal well-being.

CHAPTER 61 | METALLURGICAL APPLICATIONS IN THE ENERGY SECTOR: CURRENT AND FUTURE DEVELOPMENTS

The energy sector plays a crucial role in global sustainability and economic growth. This chapter explores the diverse metallurgical applications in the energy sector, including traditional and renewable energy sources. It highlights current practices, technological advancements, and emerging trends that shape the industry's present and future.

Metallurgical Challenges in the Energy Sector

The energy sector presents unique challenges for metallurgy, including high-temperature environments, corrosive conditions, and mechanical stress. This section discusses the specific metallurgical challenges encountered in fossil fuel-based power generation, nuclear energy, and renewable energy systems.

Metallurgy in Fossil Fuel-Based Power Generation

Fossil fuel-based power plants, such as coal-fired and natural gas plants, rely on metallurgical components for their operation. This section explores the metallurgical applications in boilers, turbines, and heat exchangers, emphasizing the need for materials with high temperature and corrosion resistance to ensure efficient and reliable power generation.

Metallurgy in Nuclear Energy

Nuclear power plants require specialized metallurgical solutions to withstand extreme conditions, including high radiation levels and elevated temperatures. This section discusses the materials used in nuclear reactors, fuel fabrication, and waste management, focusing on safety, reliability, and long-term performance.

Metallurgical Applications in Renewable Energy

Renewable energy sources, such as solar, wind, and hydroelectric power, also rely on metallurgy for the development of efficient and sustainable energy systems. This section explores the role of metallurgy in solar panels, wind turbines, energy storage systems, and hydropower infrastructure, highlighting the need for lightweight, durable, and corrosion-resistant materials.

Advancements in Metallurgy for Energy Efficiency

Efficiency is a key consideration in the energy sector. This section discusses the role of metallurgy in improving energy efficiency through materials development, such as advanced alloys, coatings, and composites. It explores the use of high-temperature materials, thermal barrier coatings, and heat transfer enhancement techniques to optimize energy conversion processes.

Materials for Energy Storage and Conversion

Energy storage and conversion technologies, such as batteries and fuel cells, require specialized materials to enable efficient and reliable energy storage and conversion. This section examines the metallurgical advancements in electrode materials, electrolytes, and catalysts, contributing to the development of high-performance energy storage and conversion systems.

Emerging Trends in Metallurgical Applications

The energy sector is evolving rapidly, driven by technological advancements and the growing demand for sustainable energy solutions. This section explores emerging trends, such as additive manufacturing of energy components, nanostructured materials for enhanced performance, and materials for next-generation energy systems.

Environmental Considerations and Sustainability

The energy sector's environmental impact is a significant concern, necessitating sustainable metallurgical practices. This section discusses the importance of materials recycling, waste management, and reducing the carbon footprint of energy production and consumption. It also explores the development of sustainable materials for energy applications.

Cross-Sector Collaborations and Research Initiatives

Addressing the metallurgical challenges in the energy sector requires collaboration among industry stakeholders, research institutions, and policymakers. This section highlights the importance of cross-sector collaborations, research initiatives, and knowledge sharing platforms in advancing metallurgical applications in the energy sector.

Future Outlook and Opportunities

The energy sector is undergoing a transformative phase, driven by decarbonization efforts, technological innovations, and shifting energy landscapes. This section provides a future outlook on the role of metallurgy in the energy sector, discussing opportunities for materials development, process optimization, and sustainable practices in line with the evolving energy needs.

Conclusion

Metallurgical applications in the energy sector are critical for ensuring efficient, reliable, and sustainable energy production and utilization. The ongoing advancements in

Chapter 62 | Heat and Mass Balance in Metallurgy

In metallurgical processes, heat and mass balance are essential considerations for optimizing process efficiency, controlling product quality, and ensuring the safety and sustainability of operations. This chapter explores the principles, methodologies, and applications of heat and mass balance in metallurgy, providing insights into their importance and practical implementation.

1. Fundamentals of Heat and Mass Transfer

This section provides an overview of the fundamental principles of heat and mass transfer, including conduction, convection, and radiation. It explains how these principles apply to metallurgical processes, where the transfer of heat and mass occurs through various mechanisms and across different phases.

2. Heat Balance in Metallurgical Processes

Heat balance involves quantifying the heat inputs and outputs within a metallurgical system to maintain thermal equilibrium. This section discusses the components of a heat balance equation, including heat sources, heat sinks, heat losses, and heat transfer mechanisms. It explores the challenges and considerations specific to metallurgical processes, such as high temperatures, phase changes, and heat recovery systems.

3. Mass Balance in Metallurgical Processes

Mass balance involves accounting for the inflow and outflow of materials in a metallurgical system. This section explains the principles of mass conservation and discusses the importance of accurate mass balance calculations in ensuring product quality, process efficiency, and environmental compliance. It covers topics such as material flow diagrams, material loss prevention, and waste management strategies.

4. Heat and Mass Balance in Pyrometallurgical Processes

Pyrometallurgical processes, such as smelting and refining, involve high-temperature operations where heat and mass transfer play crucial roles. This section focuses on the application of heat and mass balance in pyrometallurgical systems, including considerations for energy consumption, heat recovery, gas emissions, and material conversion efficiency.

5. Heat and Mass Balance in Hydrometallurgical Processes

Hydrometallurgical processes, which involve the use of aqueous solutions for metal extraction and recovery, also require careful heat and mass balance considerations. This section explores the unique challenges associated with hydrometallurgical systems, including water usage, solution chemistry, heat exchange, and solvent extraction. It discusses strategies for optimizing energy efficiency and minimizing environmental impact.

6. Process Optimization through Heat and Mass Balance

Heat and mass balance calculations provide valuable insights for process optimization and troubleshooting in metallurgical operations. This section discusses the use of heat and mass balance data to identify energy and material losses, optimize process parameters, and improve overall system performance. It explores techniques such as pinch analysis, exergy analysis, and computational modeling for process optimization.

7. Integration of Heat and Mass Balance with Process Control

Effective process control relies on accurate heat and mass balance information. This section explores the integration of heat and mass balance calculations with process control systems, highlighting the importance of real-time data acquisition, monitoring, and feedback mechanisms. It discusses the role of advanced control strategies, such as model predictive control and optimization algorithms, in enhancing process performance.

8. Environmental Implications and Sustainability

Heat and mass balance considerations are closely linked to environmental sustainability in metallurgical processes. This section discusses the environmental implications of energy consumption, greenhouse gas emissions, and material waste in metallurgical operations. It explores strategies for improving sustainability, including energy recovery, waste heat utilization, and circular economy approaches.

Future Trends and Outlook

The field of heat and mass balance in metallurgy continues to evolve, driven by advancements in technology, increasing environmental consciousness, and the pursuit of energy efficiency. This section explores the future trends and potential developments in heat and mass balance in metallurgy, including:

a) Integration of advanced sensors and automation: The use of advanced sensors and automation technologies will enable real-time data collection and analysis, facilitating more accurate heat and mass balance calculations. This integration will enhance process control, improve efficiency, and enable proactive decision-making.

b) Adoption of digitalization and artificial intelligence: The application of digitalization, data analytics, and artificial intelligence techniques will revolutionize heat and mass balance calculations. These technologies can handle large datasets, identify patterns, and optimize processes, leading to more precise heat and mass balance assessments and improved process performance.

c) Focus on circular economy and waste reduction: Metallurgical processes will increasingly emphasize waste reduction, resource recovery, and the circular economy principles. Heat and mass balance calculations will play a vital role in identifying opportunities for waste heat utilization, material recycling, and minimizing environmental impact.

d) Integration of renewable energy sources: The integration of renewable energy sources, such as solar and wind, in metallurgical processes will require careful heat and mass balance considerations. Balancing the intermittent nature of renewable energy inputs with the energy demands of metallurgical operations will be a key challenge and opportunity in the future.

e) Continued development of process modeling and simulation: The use of advanced process modeling and simulation tools will allow for more accurate predictions of heat and mass transfer phenomena in metallurgical processes. This will facilitate process optimization, equipment design improvements, and energy-efficient operations.

Conclusion

Heat and mass balance are fundamental principles in metallurgical processes, enabling efficient and sustainable operations. The comprehensive understanding and application of heat and mass balance in metallurgy contribute to improved process efficiency, reduced energy consumption, minimized environmental impact, and enhanced product quality. By embracing the evolving technologies and implementing best practices, the metallurgical industry can optimize its operations and contribute to a more sustainable future.

Chapter 63 | The Use of Artificial Intelligence in Metallurgical Research and Industry

Artificial Intelligence (AI) has emerged as a transformative technology in various fields, and its application in metallurgical research and industry holds immense potential. This chapter explores the use of AI techniques, including machine learning, deep learning, and data analytics, in advancing metallurgical research, process optimization, and quality control.

Fundamentals of Artificial Intelligence

This section provides an introduction to the fundamental concepts of AI, including machine learning algorithms, neural networks, and data preprocessing techniques. It discusses the different types of AI approaches used in metallurgy and their applicability to solving complex metallurgical challenges.

Machine Learning in Metallurgical Research

Machine learning algorithms have the ability to analyze large volumes of data and identify patterns, enabling better understanding of metallurgical phenomena. This section explores the use of machine learning techniques in metallurgical research, including material characterization, phase transformation prediction, and property modeling. It highlights the benefits of AI in accelerating research and facilitating data-driven decision-making.

AI-Driven Process Optimization

AI techniques offer opportunities for optimizing metallurgical processes, improving efficiency, and reducing energy consumption. This section discusses the use of AI in process modeling, control, and optimization. It explores applications such as intelligent process monitoring, adaptive control systems, and real-time decision-making based on AI algorithms. The potential for AI to enhance process efficiency and product quality is emphasized.

Quality Control and Defect Detection

AI-based systems can significantly enhance quality control processes in metallurgical production. This section explores the use of AI for defect detection, quality assessment, and anomaly detection in manufacturing processes. It discusses the integration of AI with advanced imaging techniques, sensors, and data analysis methods to detect and classify defects accurately, leading to improved product quality and reduced scrap rates.

Predictive Maintenance and Reliability Analysis

AI techniques enable predictive maintenance and reliability analysis in metallurgical plants. This section discusses the use of AI for condition monitoring, fault detection, and remaining useful life prediction of critical equipment and components. It explores the integration of AI with sensor data, historical maintenance records, and machine learning algorithms to optimize maintenance schedules, reduce downtime, and increase overall equipment effectiveness.

Data Analytics and Knowledge Discovery

AI-driven data analytics plays a crucial role in extracting valuable insights from large datasets in metallurgical research and industry. This section discusses data preprocessing, feature engineering, and data mining techniques for extracting knowledge from metallurgical data. It highlights the importance of data integration, data quality assessment, and data-driven decision-making in metallurgical applications.

Challenges and Ethical Considerations

The use of AI in metallurgical research and industry presents certain challenges and ethical considerations. This section addresses issues such as data privacy, algorithm transparency, and bias in AI models. It discusses the need for robust validation and verification of AI models and emphasizes the importance of human expertise in guiding AI-driven processes.

Future Directions and Opportunities:

The chapter concludes by highlighting future directions and opportunities for the use of AI in metallurgical research and industry. It discusses emerging trends, such as the integration of AI with other technologies like robotics and the Internet of Things (IoT). It also explores the potential for AI to drive innovations in materials design, process optimization, and sustainability in the metallurgical field.

Conclusion

The adoption of AI techniques in metallurgical research and industry has the potential to revolutionize processes, enhance product quality, and improve operational efficiency. The integration of machine learning, data analytics, and predictive modeling enables metallurgists to gain valuable insights, optimize processes, and make informed decisions. By embracing AI-driven approaches, the metallurgical community can unlock new opportunities and drive advancements in materials science, process optimization, and sustainable metallurgical practices.

CHAPTER 64 | UNDERSTANDING THE METALLURGICAL PROPERTIES OF NON-FERROUS METALS

Non-ferrous metals play a crucial role in various industries due to their unique properties and wide range of applications. This chapter focuses on providing a comprehensive understanding of the metallurgical properties of non-ferrous metals, including their structure, mechanical behavior, corrosion resistance, and processing characteristics.

Overview of Non-Ferrous Metals

This section introduces non-ferrous metals and discusses their classification, distinguishing them from ferrous metals. It provides an overview of commonly encountered non-ferrous metals, such as aluminum, copper, zinc, nickel, titanium, and their alloys, highlighting their significance in different industries.

Crystal Structure and Phase Transformations

The crystal structure of non-ferrous metals significantly influences their properties. This section explores the crystal structures and phase transformations observed in non-ferrous metals, explaining their impact on mechanical strength, ductility, and other mechanical properties. It discusses solid solution formation, precipitation hardening, and other phase transformations.

Mechanical Properties

Understanding the mechanical properties of non-ferrous metals is essential for their successful application. This section delves into topics such as tensile strength, yield strength, hardness, impact resistance, and fatigue behavior of non-ferrous metals. It discusses the factors influencing these properties, including alloy composition, microstructure, and processing techniques.

Corrosion Resistance

Non-ferrous metals are often chosen for their superior corrosion resistance compared to ferrous metals. This section explores the corrosion behavior of non-ferrous metals, including factors affecting corrosion resistance, types of corrosion mechanisms, and protective measures. It discusses the role of alloying elements and surface treatments in enhancing corrosion resistance.

Processing and Forming

The processing and forming techniques for non-ferrous metals differ from those used for ferrous metals. This section covers topics such as casting, rolling, extrusion, forging, and machining of non-ferrous metals. It explores the effects of processing parameters on microstructure and mechanical properties, as well as the challenges and considerations specific to non-ferrous metal processing.

Joining and Welding

Joining non-ferrous metals poses unique challenges due to their different melting points, thermal conductivity, and metallurgical behavior. This section discusses various joining techniques for non-ferrous metals, such as welding, brazing, and soldering. It covers the selection of appropriate filler materials, joint design considerations, and the impact of welding parameters on joint integrity.

Heat Treatment and Alloy Design

Heat treatment plays a significant role in optimizing the properties of non-ferrous metals and their alloys. This section explores the heat treatment techniques used for non-ferrous metals, including annealing, solution treatment, precipitation hardening, and quenching. It discusses the effects of heat treatment on microstructure, mechanical properties, and alloy design.

Applications

There are practical applications of non-ferrous metals in various industries such as aerospace, automotive, electrical, and construction sectors, showcasing how the understanding of metallurgical properties enables the selection of appropriate materials and the design of reliable components.

Conclusion

Understanding the metallurgical properties of non-ferrous metals is crucial for their successful utilization in different industries. This chapter provides a comprehensive overview of the crystal structure, mechanical behavior, corrosion resistance, processing techniques, and alloy design principles specific to non-ferrous metals. By gaining insights into these properties, engineers and metallurgists can make informed decisions regarding material selection, processing techniques, and design considerations for non-ferrous metal applications, promoting innovation and efficiency in various industries.

CHAPTER 65 | MATERIAL SELECTION IN ENGINEERING APPLICATIONS

Material selection is a crucial aspect of engineering design, where the appropriate choice of materials directly impacts the performance, reliability, and safety of engineered products. This chapter provides a comprehensive overview of material selection in engineering applications, covering key considerations, methodologies, and tools used in the process. It explores the significance of material properties, design requirements, environmental factors, and emerging trends in material selection.

Importance of Material Selection

This section emphasizes the fundamental importance of material selection in engineering applications. It discusses how material properties and characteristics directly influence the functionality, durability, and cost-effectiveness of engineered products. The section highlights the need to consider multiple factors, including mechanical properties, thermal behavior, chemical compatibility, manufacturability, and cost, to make informed material selection decisions.

Material Properties and Performance

This section delves into the various material properties that influence material selection. It provides an in-depth exploration of mechanical properties such as strength, stiffness, toughness, and fatigue resistance, as well as thermal properties including thermal conductivity, coefficient of thermal expansion, and heat resistance. Other important properties such as electrical conductivity, corrosion resistance, wear resistance, and chemical compatibility are also discussed, highlighting their significance in determining material performance.

Design Requirements and Constraints

Material selection is driven by specific design requirements and constraints imposed by engineering applications. This section examines various design factors that impact material selection decisions, such as load-bearing capacity, dimensional stability, environmental exposure, structural integrity, and aesthetic considerations. It discusses the importance of understanding the design requirements in detail and how they guide the material selection process.

Material Selection Methods and Tools

This section presents different methodologies and tools employed in material selection. It discusses traditional approaches, such as handbooks, material databases, and material property charts, which aid in preliminary material screening and selection. The section also explores computer-aided methods, including computer-aided design (CAD) and computer-aided engineering (CAE) software, finite element analysis (FEA), and optimization algorithms, which provide more advanced material selection capabilities.

Environmental and Sustainability Considerations

With growing environmental concerns, this section highlights the importance of considering the environmental impact and sustainability of materials. It discusses life cycle assessment (LCA) as a tool for evaluating the environmental footprint of materials throughout their entire life cycle. Additionally, it explores the concept of eco-design and the selection of eco-friendly materials, focusing on aspects such as recyclability, energy efficiency, and reduction of hazardous substances.

Material Selection Case Studies

This section presents a series of case studies that illustrate the practical application of material selection principles in engineering projects. Each case study highlights the challenges faced, the specific design requirements, the considered materials, and the ultimate material selection decision. The case studies cover a wide range of applications, including automotive components, aerospace structures, consumer electronics, and infrastructure projects.

Emerging Trends and Future Directions

This section discusses emerging trends and future directions in material selection. It explores the integration of advanced materials, such as composites, nanomaterials, and biomaterials, into engineering applications. It also covers the impact of additive manufacturing (3D printing) on material selection, as well as the growing interest in sustainable materials and circular economy principles. The section emphasizes the importance of staying updated with advancements in materials science and engineering to make informed material selection decisions.

Conclusion

Material selection is a critical process in engineering design, influencing the performance, reliability, and sustainability of engineered products. This chapter has provided a detailed examination of material selection in engineering applications, covering the importance of material properties, design requirements, environmental considerations, and emerging trends. By employing systematic methodologies and utilizing advanced tools, engineers can make informed material selection decisions that optimize performance, meet design objectives, and align with

CHAPTER 66 | THE IMPACT OF METALLURGICAL PROCESSES ON THE ENVIRONMENT AND PUBLIC HEALTH

Metallurgical processes play a crucial role in various industries, including mining, metal production, and manufacturing. While these processes are essential for economic development, they can also have significant environmental and public health implications. This chapter explores the impact of metallurgical processes on the environment and public health, discussing key pollutants, health risks, and mitigation strategies.

Metallurgical Processes and Environmental Impact

This section provides an overview of common metallurgical processes and their associated environmental impacts. It discusses mining activities and the extraction of raw materials, highlighting the destruction of ecosystems, habitat degradation, and soil erosion. The section also explores the emissions and waste generated during metal refining, casting, and fabrication, including air pollutants, water contamination, and solid waste disposal.

Air Pollution from Metallurgical Processes

Metallurgical processes often release various air pollutants that can have detrimental effects on air quality and human health. This section focuses on emissions such as particulate matter (PM), sulfur dioxide (SO2), nitrogen oxides (NOx), volatile organic compounds (VOCs), and heavy metals. It discusses their sources, dispersion patterns, and potential health impacts, including respiratory problems, cardiovascular diseases, and the formation of smog.

Water Contamination and Treatment

Metallurgical processes can contaminate water sources through the discharge of process wastewater, mine tailings, and chemical leaching. This section examines the pollutants commonly found in water, such as heavy metals, acids, and suspended solids, and their potential effects on aquatic ecosystems and human health. It also discusses the importance of water treatment technologies and sustainable management practices to mitigate water pollution.

Soil and Land Contamination

Metallurgical processes can result in soil and land contamination through the deposition of waste materials, spills, and leakage from storage facilities. This section explores the impact of heavy metals, acidic compounds, and other contaminants on soil quality, nutrient cycles, and plant growth. It discusses the potential for soil erosion, loss of agricultural productivity, and long-term ecological consequences. Strategies for remediation and land reclamation are also discussed.

Health Risks to Workers and Communities

Metallurgical processes can pose significant health risks to workers directly involved in these industries and nearby communities. This section examines the occupational hazards associated with exposure to harmful substances, including heavy metals, carcinogens, and particulate matter. It discusses the potential health effects, such as respiratory diseases, cancer, and neurological disorders, and emphasizes the importance of occupational safety measures and monitoring.

Regulatory Framework and Environmental Management

To mitigate the environmental and health impacts of metallurgical processes, this section discusses the importance of regulatory frameworks and environmental management practices. It examines the role of government regulations, industry standards, and international agreements in enforcing environmental protection measures. The section also explores the concept of sustainable mining and metallurgical practices, including cleaner production techniques, waste minimization, and energy efficiency.

Emerging Technologies and Best Practices

This section highlights emerging technologies and best practices aimed at reducing the environmental footprint of metallurgical processes. It discusses advancements in process optimization, recycling and reuse of materials, and the use of renewable energy sources. Additionally, it explores the concept of circular economy principles, emphasizing the importance of resource efficiency and sustainable material management.

Conclusion

Metallurgical processes have significant implications for the environment and public health. Understanding and mitigating the environmental impact of these processes is crucial for sustainable development. This chapter has examined the various environmental pollutants, health risks, and mitigation strategies associated with metallurgical processes. By implementing stringent regulations, adopting cleaner production techniques, and promoting sustainable practices, the industry can minimize its environmental footprint and safeguard public health.

CHAPTER 67 | CONCEPT OF PHASE TRANSFORMATION

Phase transformation is a fundamental concept in metallurgy that describes the change in the microstructure and properties of a material as it undergoes a transition from one phase to another. This chapter provides a comprehensive overview of the concept of phase transformation, its underlying principles, and its significance in metallurgical processes and materials development.

Fundamentals of Phase Transformation

This section introduces the basic principles of phase transformation, including thermodynamics and kinetics. It explains the concept of phases, the equilibrium phase diagram, and the driving forces that govern phase transformations. Key terms such as solid-state, diffusion-driven, and martensitic transformations are defined and discussed.

Phase Diagrams

Phase diagrams play a crucial role in understanding and predicting phase transformations in materials. This section explains the construction and interpretation of phase diagrams, including binary and ternary systems. It covers important features such as phase boundaries, phase equilibria, and the concept of solidification and solid-state transformations.

Nucleation and Growth

Nucleation and growth are fundamental processes that occur during phase transformations. This section explains nucleation as the formation of new phases and discusses factors influencing nucleation rates, such as temperature, composition, and impurities. It also explores the growth mechanisms of transformed phases and the influence of diffusion on growth kinetics.

Diffusion and Phase Transformations

Diffusion is a critical phenomenon that governs the rate and extent of phase transformations. This section delves into the role of diffusion in various types of phase transformations, such as solid-state diffusion, interface-controlled reactions, and diffusionless transformations. It discusses diffusion mechanisms and their impact on microstructure evolution.

Types of Phase Transformations

This section explores different types of phase transformations encountered in metallurgy, including solidification, precipitation, eutectoid, and martensitic transformations. It explains the mechanisms, microstructural changes, and associated properties for each type of transformation. Examples of phase transformations in various alloy systems are provided.

Phase Transformation Kinetics

Understanding the kinetics of phase transformations is essential for controlling and optimizing material properties. This section discusses the factors influencing transformation kinetics, such as temperature, alloy composition, and processing conditions. It covers topics such as nucleation and growth rates, phase transformation time-temperature-transformation (TTT) diagrams, and continuous cooling transformation (CCT) diagrams.

Phase Transformation and Material Properties

Phase transformations significantly impact the mechanical, thermal, and electrical properties of materials. This section explores the relationship between phase transformations and material properties, highlighting the effects of phase composition, grain structure, and phase distribution. It discusses how phase transformations can be tailored to achieve desired material characteristics.

Engineering Applications of Phase Transformations

Phase transformations find extensive applications in various fields of engineering and technology. This section presents examples of the practical applications of phase transformations, such as heat treatment processes, alloy design, shape memory alloys, and superalloys. It discusses how understanding and controlling phase transformations contribute to improving material performance and functionality.

Conclusion

The concept of phase transformation is a fundamental pillar of metallurgy, shaping the microstructure and properties of materials. This chapter has provided an in-depth exploration of the fundamentals of phase transformation, including phase diagrams, nucleation and growth, diffusion, types of transformations, kinetics, and their impact on material properties. By grasping the principles and mechanisms of phase transformations, metallurgists can manipulate and optimize material behavior for various engineering applications. Continued research in phase transformation holds immense potential for advancing the field of metallurgy and unlocking new possibilities in materials science.

CHAPTER 68 | BEST PRACTICES FOR THE PRODUCTION AND QUALITY CONTROL OF ADVANCED HIGH-STRENGTH STEELS

Advanced high-strength steels (AHSS) have revolutionized the automotive, aerospace, and construction industries by providing enhanced strength, lightweight properties, and improved performance. This chapter focuses on the best practices for the production and quality control of AHSS, emphasizing the key considerations and processes involved in achieving high-quality steel products.

Understanding Advanced High-Strength Steels

This section provides an overview of AHSS, discussing their classification, composition, and mechanical properties. It explains the importance of AHSS in meeting the growing demands for lightweight materials and high strength-to-weight ratios. The section also highlights the different types of AHSS, including dual-phase steels, transformation-induced plasticity steels, and martensitic steels.

Raw Material Selection and Handling

The selection of high-quality raw materials is critical for producing AHSS with consistent properties. This section discusses the criteria for selecting appropriate steel grades, including chemical composition, microalloying elements, and clean steel production techniques. It also covers the handling and storage practices to minimize contamination and maintain material integrity.

Steelmaking Processes for AHSS

This section focuses on the steelmaking processes specific to AHSS production. It discusses primary steelmaking methods such as electric arc furnace (EAF) and basic oxygen furnace (BOF), as well as secondary refining processes. The section highlights the importance of controlling alloying elements, optimizing temperature and composition, and employing advanced techniques like vacuum degassing and ladle metallurgy.

Hot Rolling and Thermomechanical Processing

Hot rolling and thermomechanical processing are crucial steps in the production of AHSS. This section explores the best practices for controlling rolling parameters, including temperature, reduction ratios, and cooling rates, to achieve desired microstructures and mechanical properties. It also discusses the use of accelerated cooling, quenching, and controlled cooling techniques to promote phase transformations and refine the microstructure.

Heat Treatment and Annealing

Heat treatment and annealing processes play a vital role in optimizing the mechanical properties and final microstructure of AHSS. This section examines the various heat treatment techniques, such as quenching and tempering, martensite tempering, and annealing, along with the associated parameters, time-temperature profiles, and cooling methods. It emphasizes the importance of precise control to achieve the desired balance of strength, toughness, and formability.

Quality Control and Testing Methods

Ensuring the quality and consistency of AHSS requires rigorous quality control measures and testing methods. This section discusses the best practices for quality control at different stages of production, including raw material inspection, in-process monitoring, and final product testing. It covers mechanical testing, non-destructive testing, metallographic analysis, and chemical analysis techniques to assess the mechanical properties, microstructure, and chemical composition of

AHSS.

Continuous Improvement and Lean Manufacturing

Continuous improvement and lean manufacturing principles are essential for optimizing the production of AHSS. This section explores strategies for process optimization, waste reduction, and productivity enhancement through the adoption of lean manufacturing principles. It emphasizes the importance of data-driven decision-making, employee engagement, and the implementation of quality management systems to drive continuous improvement.

Environmental and Sustainability Considerations

While producing AHSS, it is essential to consider environmental and sustainability aspects. This section discusses the implementation of energy-efficient practices, waste management strategies, and the reduction of greenhouse gas emissions in AHSS production. It also explores the concept of life cycle assessment to evaluate the environmental impact of AHSS and promotes the use of recycled materials and circular economy principles.

Conclusion

The production of advanced high-strength steels requires a comprehensive understanding of the material properties, careful selection of raw materials, and adherence to stringent production and quality control practices. This chapter has provided an in-depth exploration of the best practices for the production and quality control of AHSS. By following these practices, manufacturers can ensure the consistent production of high-quality AHSS with desired mechanical properties and microstructures.

It is important to emphasize the continuous improvement and lean manufacturing principles to optimize processes, reduce waste, and enhance productivity. Additionally, the consideration of environmental and sustainability aspects is crucial in minimizing the environmental impact of AHSS production and promoting sustainable practices.

By implementing these best practices, manufacturers can meet the increasing demand for advanced high-strength steels in various engineering applications. AHSS offers significant advantages in terms of strength, lightweight properties, and performance, making them invaluable materials for industries such as automotive, aerospace, and construction.

In conclusion, this chapter has highlighted the importance of adopting best practices for the production and quality control of advanced high-strength steels. By integrating these practices into their manufacturing processes, companies can ensure the consistent production of high-quality AHSS, meet customer requirements, and contribute to the advancement of engineering applications that rely on these exceptional materials.

Chapter 69 | The Challenges and Opportunities of Recycling Metals for Sustainable Development

The recycling of metals plays a crucial role in achieving sustainable development by conserving natural resources, reducing energy consumption, and minimizing environmental impacts. This chapter explores the challenges and opportunities associated with recycling metals, highlighting the importance of efficient recycling processes and their impact on sustainable development.

Importance of Metal Recycling

This section emphasizes the significance of metal recycling as a sustainable solution for reducing the reliance on primary metal extraction. It discusses the environmental benefits of metal recycling, including energy savings, reduced greenhouse gas emissions, and conservation of natural resources. The section also addresses the growing demand for recycled metals in various industries.

Challenges in Metal Recycling

Metal recycling faces several challenges that need to be addressed for sustainable development. This section examines the technical challenges, such as sorting and separation of metal alloys, removal of contaminants, and recovering metals from complex waste streams. It also explores economic challenges, including fluctuating metal prices and the cost-effectiveness of recycling processes. Furthermore, the section discusses the logistical challenges of collecting, transporting, and processing metal scrap.

Recycling Technologies and Processes

This section delves into various recycling technologies and processes employed in metal recycling. It explores mechanical recycling methods, such as shredding, sorting, and separation techniques. It also discusses pyrometallurgical processes like smelting and refining, as well as hydrometallurgical processes like leaching and solvent extraction. The section highlights advancements in recycling technologies and their potential for improving efficiency and metal recovery rates.

Innovations in Metal Recycling

This section focuses on the innovative approaches and technologies being developed to address the challenges of metal recycling. It highlights advancements in sensor-based sorting systems, intelligent automation, and robotics for efficient material separation. Additionally, it explores novel processes like electrochemical and biotechnological methods for metal recovery from electronic waste and other complex sources. The section also discusses the role of research and development in driving innovation in metal recycling.

Policy and Regulatory Framework

Effective policies and regulations are essential for promoting metal recycling and achieving sustainable development goals. This section examines the policy measures and regulatory frameworks implemented at local, national, and international levels to incentivize recycling practices, set recycling targets, and promote the circular economy. It also addresses the importance of collaboration between governments, industry stakeholders, and research institutions in shaping effective policies.

Circular Economy and Metal Recycling

The concept of the circular economy is gaining prominence in the pursuit of sustainable development. This section explores the link between metal recycling and the circular economy, emphasizing the need for a closed-loop approach where materials are recycled, reprocessed, and reused. It discusses the role of extended producer responsibility, product design for recyclability, and fostering a culture of recycling in achieving a circular economy.

Socio-economic Impacts and Opportunities

Metal recycling not only contributes to environmental sustainability but also generates socio-economic benefits. This section examines the positive impacts of metal recycling on job creation, resource conservation, and economic growth. It also explores the potential opportunities for entrepreneurship and innovation in the recycling industry, including the development of new business models and value-added products from recycled metals.

Education and Awareness

Education and awareness play a crucial role in promoting metal recycling and sustainable practices. This section emphasizes the need for public awareness campaigns, educational programs, and training initiatives to foster a recycling mindset and encourage responsible consumer behavior. It also highlights the importance of collaboration between educational institutions, industry associations, and governmental organizations in raising awareness about the benefits of metal recycling.

Conclusion

The challenges and opportunities of recycling metals for sustainable development are multi-faceted. Efficient recycling processes, technological innovations, supportive policies, and public engagement are vital for overcoming the challenges and maximizing the benefits of metal recycling. By embracing recycling as a core element of the circular economy

CHAPTER 70 | TESTING METHODS FOR REFRACTORY MATERIALS

Refractory materials play a critical role in various high-temperature industrial applications. Ensuring the quality and performance of refractories is essential for the success and efficiency of these processes. This chapter focuses on the testing methods employed to evaluate the properties and characteristics of refractory materials, providing insights into their suitability for specific applications.

Physical Testing Methods

Physical testing methods are used to assess the physical properties of refractory materials. This section discusses techniques such as density measurement, porosity determination, and thermal conductivity testing. It also explores the use of techniques like dilatometry for assessing the dimensional stability of refractories under thermal stress. Additionally, it covers mechanical testing methods, including compressive strength, flexural strength, and abrasion resistance tests.

Thermal conductivity testing

This testing provides valuable information about their thermal transport properties, which is crucial for assessing their performance in high-temperature applications. This section provides an overview of thermal conductivity testing methods commonly employed for refractory materials.

1. Guarded Hot Plate Method:
The guarded hot plate method is a widely used technique for measuring the thermal conductivity of refractory materials. In this method, a thin sample of the refractory material is sandwiched between two temperature-controlled plates. One plate is heated, while the other is cooled, creating a temperature gradient across the sample. By measuring the temperature difference and the heat flux passing through the sample, the thermal conductivity can be calculated using Fourier's

law of heat conduction.

2. Laser Flash Method:

The laser flash method is another commonly used technique for measuring the thermal conductivity of refractory materials. In this method, a short laser pulse is directed onto the surface of a small disc-shaped sample. The laser pulse generates a localized heat source, and the resulting temperature rise is detected using a fast-response sensor. By measuring the temperature rise as a function of time, the thermal diffusivity of the material can be determined, from which the thermal conductivity can be calculated.

3. Transient Plane Source (TPS) Method:

The transient plane source method is particularly suitable for measuring the thermal conductivity of thin and low-conductivity refractory materials. In this method, a thin disc-shaped sensor with a known heating element is placed in contact with the sample. The heating element generates a heat pulse, and the subsequent temperature rise in the sensor is recorded. By analyzing the temperature response, the thermal conductivity of the refractory material can be calculated.

4. Comparative Methods:

In some cases, comparative methods are employed to estimate the thermal conductivity of refractory materials. These methods involve comparing the thermal conductivities of the refractory material with known reference materials. For example, the hot-wire method compares the thermal conductivity of the refractory material with that of a standard wire material. Although these methods provide relative measurements, they can be useful for quick assessments or quality control purposes.

5. Factors to Consider:

When conducting thermal conductivity testing for refractory materials, several factors should be considered. These include sample preparation, ensuring uniform contact between the sample and the measuring instrument, and accounting for any anisotropic or directional properties of the material. It is also important to consider the temperature range and conditions relevant to the intended application of the refractory material.

Thermal Testing Methods

The behavior of refractory materials under high-temperature conditions is crucial for their suitability in specific applications. This section focuses on thermal testing methods used to evaluate refractories' thermal properties. It includes techniques such as thermal conductivity measurement, thermal expansion analysis, and differential thermal analysis (DTA) to understand the thermal behavior, heat transfer, and thermal stability of refractories.

Chemical Analysis Methods

Chemical composition and purity are critical factors that determine the performance and compatibility of refractory materials. This section explores chemical analysis methods, such as X-ray fluorescence (XRF) and atomic absorption spectroscopy (AAS), used to determine the elemental composition of refractories. It also discusses techniques like acid digestion and wet chemical analysis for assessing impurities and contaminants.

Microstructural Analysis Methods

The microstructure of refractory materials significantly influences their properties and performance. This section delves into microstructural analysis methods, including optical microscopy, scanning electron microscopy (SEM), and transmission electron microscopy (TEM). These techniques provide valuable insights into grain size, phase composition, porosity distribution, and the presence of any defects or cracks within the refractory structure.

Thermo-mechanical Testing Methods

Refractories are subjected to thermal and mechanical stresses during operation, which can affect their integrity and durability. This section focuses on thermo-mechanical testing methods used to evaluate the response of refractory materials under combined thermal and mechanical loading conditions. It includes techniques such as thermal shock resistance testing, hot modulus of rupture (HMOR) testing, and creep testing to assess the refractories' resistance to thermal cycling, mechanical strength, and deformation behavior.

Corrosion and Erosion Testing Methods

Refractories may be exposed to aggressive environments, including chemical corrosion and mechanical erosion. This section explores testing methods used to evaluate the resistance of refractory materials to corrosive and erosive conditions. Techniques like slag testing, acid resistance testing, and erosion testing simulate the real-world conditions and provide insights into the refractories' resistance to degradation.

Performance Testing Methods

To ensure the suitability of refractory materials for specific industrial applications, performance testing methods are employed. This section discusses techniques like thermal shock testing, thermal cycling testing, and high-temperature performance testing in real operating conditions. It emphasizes the importance of evaluating refractories' performance in the actual environment they will encounter to assess their long-term durability and reliability.

Conclusion

Testing methods for refractory materials are crucial for evaluating their physical, thermal, chemical, and mechanical properties. By employing these methods, manufacturers, researchers, and users can make informed decisions about the selection, design, and optimization of refractories for various industrial applications. The comprehensive understanding of refractory behavior obtained through testing aids in improving their performance, extending their lifespan, and ensuring the efficiency and reliability of high-temperature processes.

Chapter 71 | The Role of Thermodynamics in Metallurgy

The field of metallurgy relies heavily on the principles of thermodynamics to understand and manipulate the behavior of metals and alloys. Thermodynamics provides the foundation for studying the energetics and equilibrium of metallurgical processes, enabling the prediction and control of phase transformations, alloying behavior, and reaction kinetics. This chapter explores the fundamental concepts of thermodynamics and their application in metallurgy.

Basics of Thermodynamics

1. *Laws of Thermodynamics:* A brief overview of the laws of thermodynamics, including the zeroth law, first law, second law, and third law, and their significance in understanding energy, heat, and entropy.

2. *Thermodynamic Systems and Surroundings:* Defining and understanding the concepts of open, closed, and isolated systems, as well as the role of surroundings in energy exchange.

3. *Thermodynamic Properties:* Introduction to extensive and intensive properties, including temperature, pressure, volume, and internal energy, and their relevance to metallurgical processes.

Thermodynamics of Phase Transformations

1. *Phase Equilibria:* Understanding phase diagrams, phase boundaries, and phase transformations in metallurgical systems. Discussing binary and ternary phase diagrams, solid solutions, eutectic and peritectic reactions, and the lever rule.

2. *Gibbs Phase Rule:* Exploring the Gibbs phase rule and its

application in determining the number of degrees of freedom in a system with multiple phases.

3. ***Thermodynamic Driving Forces:*** Discussing the concept of chemical potential and its role as a driving force for phase transformations, including diffusion, nucleation, and growth.

Thermodynamics of Alloying and Solution Behavior

1. ***Ideal Solutions:*** Introducing ideal solutions and discussing their thermodynamic behavior, including Raoult's law and Henry's law.

2. ***Non-Ideal Solutions:*** Exploring non-ideal solution behavior, including deviations from ideal solution behavior, activities, and activity coefficients.

3. ***Solid Solutions:*** Understanding the thermodynamics of solid solutions, including the regular solution model, Hume-Rothery rules, and the effect of temperature, composition, and atomic size on solid solubility.

Thermodynamics of Reactions and Equilibrium

1. ***Chemical Reactions:*** Investigating the thermodynamics of chemical reactions, including the calculation of reaction Gibbs energy, equilibrium constants, and the relationship between equilibrium constants and temperature.

2. ***Ellingham Diagrams:*** Discussing the use of Ellingham diagrams to analyze the thermodynamics of high-temperature reactions, such as oxidation, reduction, and formation of compounds.

3. ***Equilibrium Calculations:*** Exploring equilibrium calculations using thermodynamic data, including the calculation of equilibrium compositions, equilibrium constant determination, and the effect of temperature and pressure on equilibrium.

Thermodynamics in Process Optimization and Design

1. *Energy and Heat Balance:* Understanding the importance of energy and heat balance in metallurgical processes, including heat transfer, heat exchangers, and energy conservation.

2. *Reaction Kinetics:* Discussing the interplay between thermodynamics and reaction kinetics in controlling the rate of metallurgical processes, including the Arrhenius equation, activation energy, and reaction mechanisms.

3. *Process Modeling and Simulation:* Exploring the use of thermodynamics in process modeling and simulation, including the development of phase equilibria models and computational tools for process optimization.

Conclusion

Thermodynamics is a powerful tool in metallurgy that provides a systematic framework for understanding the behavior of metals and alloys. By applying thermodynamic principles, metallurgists can predict and control phase transformations, design alloys with specific properties, optimize process conditions, and develop innovative materials. Understanding the role of thermodynamics in metallurgy is essential for advancing the field and

CHAPTER 72 | MINERAL PROCESSING AND BENEFICIATION

Mineral processing and beneficiation are essential steps in extracting valuable minerals from their ores and maximizing their economic value. This chapter provides an overview of the principles, techniques, and processes involved in mineral processing and beneficiation, highlighting their significance in the mining industry.

Introduction to Mineral Processing

1. Importance of Mineral Processing: Exploring the importance of mineral processing in unlocking the economic potential of mineral resources and meeting the demand for various raw materials.

2. Mineral Processing Objectives: Discussing the primary objectives of mineral processing, including the liberation of valuable minerals, separation of minerals from gangue, and the production of concentrates.

3. Mineral Processing Circuit: Introducing the concept of a mineral processing circuit and the various stages involved, such as comminution, sizing, concentration, and dewatering.

Comminution

1. Crushing: Exploring the purpose and methods of crushing, including primary, secondary, and tertiary crushing, as well as factors influencing the selection of crushing equipment.

2. Grinding: Discussing the principles of grinding and the different types of grinding mills used in mineral processing, including ball mills, rod mills, and autogenous mills.

3. Comminution Efficiency: Understanding the factors affecting comminution efficiency, such as particle size distribution, ore hardness, and energy consumption, and strategies for improving efficiency.

Mineral Liberation and Separation

1. Particle Size Analysis: Exploring the importance of particle size analysis in mineral processing and the techniques used, including sieving, sedimentation, and laser diffraction.

2. Mineral Liberation: Understanding the concept of mineral liberation and its significance in determining the efficiency of subsequent separation processes.

3. Gravity Separation: Discussing gravity-based separation methods, such as dense media separation, jigging, and spirals, for separating minerals based on their density differences.

4. Froth Flotation: Exploring the principles and techniques of froth flotation, a widely used method for separating valuable minerals from their associated gangue minerals based on their hydrophobicity.

5. Magnetic and Electrostatic Separation: Introducing magnetic and electrostatic separation methods for separating minerals based on their magnetic or electrical properties.

Concentration and Enrichment

1. Gravity Concentration: Discussing gravity-based concentration techniques, such as centrifugal concentrators, shaking tables, and enhanced gravity separators, for enriching valuable minerals.

2. Froth Flotation Concentration: Exploring the optimization of froth flotation processes for achieving high-grade concentrates and minimizing the loss of valuable minerals.

3. Hydrometallurgical Techniques: Introducing hydrometallurgical techniques, such as leaching and solvent extraction, for extracting valuable metals from concentrates.

Dewatering and Tailings Management

1. Dewatering Techniques: Discussing dewatering methods, such as filtration, thickening, and drying, for reducing the moisture content of mineral concentrates.

2. Tailings Management: Exploring strategies for the safe storage and management of mineral processing tailings to minimize environmental impact and ensure long-term stability.

Process Optimization and Control

1. Process Modeling and Simulation: Discussing the use of process modeling and simulation tools to optimize mineral processing operations, improve efficiency, and reduce costs.

2. Automation and Control Systems: Exploring the role of automation and control systems in monitoring and optimizing mineral processing processes, ensuring consistent product quality, and enhancing safety.

The continuous advancement and innovation in mineral processing and beneficiation techniques are crucial for meeting the ever-increasing demand for minerals while addressing sustainability challenges. Key areas of future development in mineral processing and beneficiation include:

1. Advanced Separation Techniques: Exploring new separation technologies and methods that can enhance the efficiency and selectivity of mineral separation, such as advanced froth flotation techniques, high-gradient magnetic separation, and electrostatic separation.

2. Sustainable Processing Solutions: Developing environmentally friendly and energy-efficient processing methods that minimize water and energy consumption, reduce waste generation, and optimize resource utilization. This includes the development of green reagents, water recycling systems, and process optimization tools.

3. Sensor-Based Sorting: Investigating the use of sensor-based sorting technologies, such as X-ray transmission sorting, optical sorting, and electromagnetic sensing, for efficient pre-concentration of ores, reducing the energy-intensive grinding and beneficiation steps.

4. Bioleaching and Biohydrometallurgy: Exploring the potential of bioleaching and biohydrometallurgical processes that utilize microorganisms to extract metals from low-grade ores or complex mineral matrices, reducing the need for traditional chemical processes.

5. Data Analytics and Machine Learning: Leveraging the power of data analytics and machine learning algorithms to analyze vast amounts of process data, optimize process parameters, predict performance, and enable real-time process control and decision-making.

6. Circular Economy Approaches: Embracing circular economy principles in mineral processing and beneficiation by focusing on recycling, reusing, and recovering valuable materials from waste streams, as well as exploring new ways to extract valuable minerals from secondary sources.

7. Process Intensification: Investigating novel process intensification techniques that can enhance process efficiency, reduce footprint, and increase throughput, such as microwave-assisted processing, ultrasound-assisted processing, and high-pressure grinding rolls.

8. Integration of Digital Technologies: Integrating digital technologies, such as Internet of Things (IoT), artificial intelligence (AI), and digital twins, into mineral processing operations to enable real-time monitoring, predictive maintenance, and optimization of equipment performance.

9. Tailings Valorization: Exploring innovative approaches for the valorization of mineral processing tailings, such as the recovery of valuable metals, extraction of residual mineral content, and utilization in construction materials or backfilling.

10. Collaborative Research and Industry Partnerships: Encouraging collaboration between researchers, industry stakeholders, and government agencies to foster knowledge exchange, promote research and development, and address common challenges in mineral processing and beneficiation.

By embracing these advancements and adopting best practices, the mineral processing and beneficiation industry can achieve higher efficiency, lower costs, and reduced environmental impact while ensuring the sustainable supply of minerals to meet the needs of various industries and contribute to global development.

And the successful implementation of these advancements in mineral processing and beneficiation relies on several key factors:

1. Research and Development: Continued investment in research and development activities to explore new technologies, optimize existing processes, and overcome technical challenges in mineral processing and beneficiation.

2. Technological Readiness: Ensuring that the developed technologies are scalable, economically viable, and readily implementable in industrial settings.

3. Regulatory Framework: Establishing a robust regulatory framework that promotes sustainable practices, environmental protection, and health and safety standards in the mineral processing and beneficiation industry.

4. Skills and Expertise: Developing a skilled workforce equipped with the knowledge and expertise in advanced mineral processing techniques, automation, data analytics, and sustainable practices.

5. Industry Collaboration: Encouraging collaboration and knowledge-sharing among industry stakeholders, research institutions, and government agencies to foster innovation, address common challenges, and facilitate technology transfer.

6. Market Demand and Economic Viability: Aligning the

development of mineral processing and beneficiation technologies with market demand, economic viability, and industry requirements to ensure commercialization and adoption of these technologies.

7. Social Acceptance: Considering the social and community aspects of mineral processing and beneficiation, engaging with local communities, and addressing concerns related to environmental impact, water usage, and land reclamation.

8. Continuous Improvement: Embracing a culture of continuous improvement and operational excellence in mineral processing and beneficiation operations to optimize efficiency, reduce waste generation, and enhance overall performance.

By addressing these factors and considering the holistic aspects of mineral processing and beneficiation, the industry can achieve sustainable and responsible mineral extraction and value addition, contributing to economic development while minimizing environmental impact.

Conclusion

Mineral processing and beneficiation play a vital role in extracting valuable minerals from ores and converting them into valuable products. The processes and techniques discussed in this chapter are critical for maximizing resource utilization, achieving high recovery rates, and minimizing environmental impact.

Chapter 73 | Overview of Mechanical Metallurgy

Mechanical metallurgy is a branch of metallurgical science that focuses on the mechanical properties and behavior of metals and alloys. It deals with the study of how metals respond to external forces, such as load, stress, strain, and temperature, and how these factors influence the mechanical properties of materials. Mechanical metallurgy plays a crucial role in understanding and designing materials for various engineering applications.

The field of mechanical metallurgy encompasses several key aspects:

1. Elasticity and Plasticity: Mechanical metallurgy explores the elastic and plastic deformation behavior of metals. It examines how metals respond to applied loads and how they regain their original shape after the load is removed (elastic deformation). It also investigates the permanent deformation that occurs when the material exceeds its elastic limit (plastic deformation).

2. Strength and Hardness: Understanding the strength and hardness properties of metals is essential in mechanical metallurgy. This includes studying the mechanisms of strengthening, such as alloying, heat treatment, and grain refinement, as well as evaluating the resistance of materials to deformation and fracture.

3. Fracture Mechanics: Mechanical metallurgy delves into the science of fracture and failure analysis. It investigates the factors that lead to material fracture, including stress concentrations, defects, and material microstructure. This knowledge helps in designing materials and structures to prevent catastrophic failures.

4. Fatigue and Creep: Mechanical metallurgy examines the behavior of materials under cyclic loading (fatigue) and long-term exposure to elevated temperatures (creep). It studies the mechanisms that lead to fatigue crack initiation and propagation, as well as the time-dependent deformation and failure of materials subjected to constant stress over time.

5. Materials Testing and Characterization: Mechanical metallurgy involves various testing and characterization techniques to assess the mechanical properties of metals. This includes tensile testing, hardness testing, impact testing, and non-destructive testing methods. These techniques provide valuable information on material strength, ductility, toughness, and other mechanical properties.

6. Material Selection and Design: Mechanical metallurgy plays a critical role in material selection and design for engineering applications. It helps in identifying suitable materials based on their mechanical properties, performance requirements, and environmental conditions. It also guides the design process by considering factors like load-bearing capacity, fatigue resistance, and durability.

7. Failure Analysis and Remediation: When failures occur in engineering components or structures, mechanical metallurgy is employed to analyze and understand the causes of failure. This involves investigating factors such as material defects, improper processing, environmental degradation, or excessive loading. Based on the findings, remedial measures can be taken to prevent future failures.

Overall, mechanical metallurgy provides a fundamental understanding of the mechanical behavior of metals and alloys. It enables engineers and scientists to design materials with specific mechanical properties and to ensure the safe and reliable operation of structures and components in various industries, including automotive, aerospace, construction, energy, and manufacturing.

Strengthening Mechanisms in Mechanical Metallurgy

In mechanical metallurgy, the strength of a material refers to its ability to resist deformation and withstand applied loads. Various strengthening mechanisms are employed to enhance the strength of materials, allowing them to meet the requirements of demanding applications. These mechanisms are based on altering the microstructure and controlling the dislocation motion within the material. Here are some of the key strengthening mechanisms:

1. Solid Solution Strengthening:
- Introduction of alloying elements into the base metal, which disrupts the crystal lattice and hinders dislocation movement
- Alloying elements can either form solid solutions with the base metal or create precipitates that impede dislocation motion
- Examples include the addition of carbon in steel (forming solid solution) or precipitation hardening in aluminum alloys (forming precipitates)

2. Strain Hardening (Work Hardening):
- Applying plastic deformation to a material through processes such as cold working or rolling
- Dislocations are introduced and tangled, making it harder for them to move, resulting in increased strength and hardness
- Work-hardened materials have improved yield strength and can be further strengthened by subsequent heat treatments

3. Grain Refinement:
- Reducing the grain size of a material through processes like grain refinement by severe plastic deformation (SPD) or grain boundary engineering
- Smaller grains impede dislocation movement and result in increased strength, hardness, and resistance to deformation
- Grain refinement can be achieved through techniques such as equal channel angular pressing (ECAP) or high-pressure torsion (HPT)

4. Precipitation Hardening:
- Formation of precipitates within a material matrix through controlled heat treatment processes
- Precipitates act as obstacles for dislocation movement, increasing the material's strength and hardness
- Commonly used in alloys such as aluminum, copper, and nickel-based alloys, where the precipitation of fine particles contributes to strengthening

5. Transformation Hardening:

- Inducing phase transformations in a material through heat treatment processes, such as quenching and tempering
- Transformation from one crystal structure to another can lead to the formation of new phases with improved strength and hardness
- Examples include the formation of martensite in steels or the precipitation of secondary phases in superalloys

6. Texture Hardening:
- Controlling the crystallographic orientation or texture of a material to enhance its strength
- Certain crystallographic orientations can impede dislocation movement, resulting in higher strength in specific directions
- Texture control can be achieved through processes like rolling, annealing, or severe plastic deformation techniques

7. Dispersion Strengthening:
- Introduction of fine particles or fibers into a material matrix to hinder dislocation movement and increase strength
- The dispersed particles act as barriers, impeding dislocation motion and leading to enhanced strength and hardness
- Examples include the addition of ceramic particles to metal matrices in metal matrix composites (MMCs)

Understanding these strengthening mechanisms allows engineers and metallurgists to tailor materials with desired mechanical properties. By carefully selecting and optimizing these mechanisms, materials can be developed to meet specific requirements in industries such as aerospace, automotive, construction, and many others. Continuous research and development in the field of mechanical metallurgy are essential for exploring novel strengthening mechanisms and further improving material performance.

Chapter 74 | Welding and Joining of Metals

In the field of metallurgy, welding and joining play a crucial role in fabricating structures and components by combining various metal pieces together. Welding and joining of metals is a fundamental process used to permanently join two or more metal parts together. It involves the application of heat, pressure, or a combination of both to create a bond between the materials, resulting in a strong and durable connection. This process is crucial in various industries, including automotive, aerospace, construction, and manufacturing.

The process of welding involves the localized heating of the metal parts to their melting point, followed by the application of a filler material (if required) to form a weld joint. The heat source can be an electric arc, laser, or flame. As the metal cools and solidifies, it forms a continuous joint that exhibits metallurgical bonding.

Joining of metals can also involve processes other than welding. These include solid-state joining techniques, such as diffusion bonding, friction welding, and explosion welding. In these methods, the metals are not melted but are joined by applying pressure and allowing atomic diffusion between the interfaces to create a bond.

The welding and joining processes have several advantages, including:

1. Strength and durability: Properly executed welds and joints can provide high strength and durability, allowing for the efficient transfer of loads between joined parts.

2. Versatility: Welding and joining techniques can be applied to a wide range of metals and alloys, including steel, aluminum, titanium, and nickel alloys, enabling the fabrication of complex structures with diverse material combinations.

3. Efficiency: Welding and joining processes are typically faster and more cost-effective compared to other methods of joining, such as mechanical fastening or adhesives.

4. Design flexibility: Welding and joining allow for greater design flexibility, as they enable the creation of intricate shapes and configurations by fusing or bonding multiple parts together.

However, there are also certain challenges and considerations associated with welding and joining:

1. Metallurgical changes: The heat applied during welding and joining can result in metallurgical changes in the materials, leading to altered microstructures and potential changes in mechanical properties. Careful selection of welding parameters and post-weld heat treatment can help mitigate these effects.

2. Weld quality and integrity: Ensuring high-quality welds and joints is crucial to maintain structural integrity. Factors such as proper joint preparation, suitable filler material selection, and adherence to welding procedures and standards are essential for achieving reliable and defect-free connections.

3. Distortion and residual stresses: Welding and joining processes can induce distortion and residual stresses in the joined parts, which may require subsequent corrective measures, such as mechanical straightening or post-weld heat treatment.

4. Material compatibility: Joining different metals or alloys can present challenges due to differences in their physical and chemical properties. Proper material selection, including consideration of composition, compatibility, and thermal expansion coefficients, is crucial to achieve compatible joints.

Advancements in welding and joining technologies continue to address these challenges and expand the capabilities of the process. Techniques such as laser welding, electron beam welding, friction stir welding, and additive manufacturing (3D printing) offer new possibilities for joining dissimilar materials, reducing distortion, and improving overall weld quality.

In summary, welding and joining of metals are essential processes in various industries, providing the means to create strong and reliable connections between metal parts. Understanding the principles, techniques, and considerations involved in these processes is crucial for successful fabrication and construction applications.

Chapter 75 | The Future of Metallurgy in India: Trends, Challenges, and Opportunities

Overview of the Current Metallurgical Landscape in India

The metallurgical industry plays a crucial role in India's economy, contributing significantly to its GDP and employment. India has a rich history in metallurgy, with evidence of early metalworking dating back to ancient times. Over the years, the country has developed a robust metallurgical landscape encompassing various sectors and industries.

1. Steel Industry: The steel industry is a cornerstone of India's metallurgical sector. India is currently the second-largest steel producer globally, with a strong focus on both primary and secondary steel production. Key players in the Indian steel industry include public and private sector companies, contributing to domestic consumption as well as exports.

2. Automotive Industry: The automotive sector is a major consumer of metallurgical products in India. The country is one of the world's largest automotive markets, and the demand for high-quality steels and alloys continues to grow. Metallurgical advancements and tailored material solutions play a vital role in meeting the stringent requirements of the automotive industry.

3. Construction Industry: India's rapid urbanization and infrastructure development have fueled the demand for metallurgical products in the construction sector. Steel and other metal-based materials are extensively used in building structures, bridges, and other infrastructure projects. Metallurgical advancements in areas such as corrosion resistance, strength, and durability are crucial for the construction industry.

4. Aerospace and Defense Industry: The aerospace and defense sectors in India are witnessing significant growth, with a focus on indigenous manufacturing and technological advancements. Metallurgical expertise is vital in developing specialized alloys, superalloys, and advanced materials for aircraft, satellites, and defense applications. India's defense industry heavily relies on metallurgical processes for materials used in armor, missiles, and other critical components.

5. Mining and Minerals Industry: India is rich in mineral resources, and the mining industry plays a pivotal role in supplying raw materials to the metallurgical sector. The extraction and processing of ores, such as iron ore, bauxite, and coal, are integral to metallurgical operations.

6. Research and Development: India has several research institutes, academic institutions, and industry-focused laboratories dedicated to metallurgical research and development. These entities conduct cutting-edge research, develop new alloys and processes, and provide technical support to the industry. Collaboration between academia, research organizations, and industry players is essential for advancing metallurgical knowledge and innovation.

7. Government Support: The Indian government has implemented various policies and initiatives to support the metallurgical sector. This includes investment incentives, infrastructure development, skill development programs, and the promotion of research and development activities. The government's focus on the "Make in India" initiative and the promotion of sustainable practices has further contributed to the growth of the metallurgical industry.

The current metallurgical landscape in India showcases a blend of traditional practices and technological advancements. The industry is continuously evolving to meet the growing demands of various sectors while addressing challenges such as sustainability, environmental impact, and global competitiveness. The collaboration between industry, academia, and the government is crucial in driving the future growth of the metallurgical sector in India.

Emerging Trends in Metallurgical Research and Development

Metallurgical research and development is a dynamic field that constantly evolves to meet the changing needs and demands of various industries. Several emerging trends are shaping the future of metallurgical research and development. These trends include:

1. Advanced Materials: Metallurgical research is focused on developing advanced materials with enhanced properties to meet the increasing demands of industries. This includes the development of high-strength steels, lightweight alloys, superalloys, and composite materials. Advanced materials offer improved mechanical properties, corrosion resistance, and thermal stability, opening up new possibilities for applications in sectors such as aerospace, automotive, and energy.

2. Additive Manufacturing: Additive manufacturing, also known as 3D printing, is revolutionizing the manufacturing industry, and metallurgy plays a vital role in this field. Metallurgical research is focused on developing new alloys and optimizing the manufacturing processes for additive manufacturing, enabling the production of complex geometries and customized components. Additive manufacturing offers advantages such as design flexibility, reduced material waste, and rapid prototyping, leading to advancements in industries like aerospace, medical, and automotive.

3. Sustainable Metallurgy: With the increasing emphasis on sustainability, metallurgical research is focused on developing environmentally friendly processes and materials. This includes the development of energy-efficient processes, waste reduction strategies, recycling techniques, and the use of renewable energy sources. Sustainable metallurgy aims to minimize the environmental impact of metallurgical operations while maintaining high-quality material production.

4. Digitalization and Data Analytics: The integration of digital technologies and data analytics is transforming metallurgical research and development. Advanced simulation tools, computational modelling, and machine learning algorithms are used to optimise processes, predict material properties, and accelerate materials discovery. Digitalization enables the analysis of vast amounts of data, leading to improved process control, better product quality, and enhanced efficiency in metallurgical operations.

5. Nanomaterials and Nanotechnology: Metallurgical research is exploring the unique properties and applications of nanomaterials. Nanotechnology allows the manipulation of materials at the atomic and molecular levels, resulting in improved mechanical, electrical, and thermal properties. Metallurgical research focuses on synthesizing and characterizing nanomaterials and understanding their behaviour in different environments. Nanotechnology finds applications in areas such as electronics, energy storage, catalysis, and biomedical devices.

6. Process Intensification: Process intensification involves optimizing existing metallurgical processes to enhance efficiency, reduce energy consumption, and increase productivity. Metallurgical research focuses on developing innovative process designs, reactor configurations, and control strategies to achieve these goals. Process intensification enables higher yields, improved product quality, and reduced environmental impact.

7. Multi-Scale Modeling: Metallurgical research incorporates multi-scale modeling techniques to understand and predict the behavior of materials at different length scales. This includes atomistic simulations, molecular dynamics, finite element analysis, and computational thermodynamics. Multi-scale modeling enables researchers to gain insights into the structure-property relationships of materials, facilitating the design of new materials with tailored properties.

These emerging trends in metallurgical research and development hold great promise for advancements in materials science and industrial applications. They contribute to the development of sustainable, high-performance materials, efficient manufacturing processes, and innovative solutions to address the challenges faced by various industries. Collaborations between academia, industry, and research institutions play a crucial role in driving these trends and shaping the future of metallurgical research and development.

Sustainable Metallurgy: Environmentally Conscious Practices

Sustainable metallurgy refers to the adoption of environmentally conscious practices in the field of metallurgical processes and operations. In India, there is an increasing focus on sustainable metallurgy to minimize the environmental impact of the industry while promoting responsible resource management. Here are some key aspects of sustainable metallurgy in India:

1. Resource Efficiency: Sustainable metallurgy aims to optimize the use of resources such as raw materials, energy, and water. This involves implementing efficient extraction techniques, recycling and reusing materials, and minimizing waste generation. In India, initiatives are being taken to improve resource efficiency through process optimization, waste heat recovery, and the adoption of cleaner technologies.

2. Emission Control: Controlling emissions is a significant aspect of sustainable metallurgy. India has implemented stringent regulations and standards to limit air and water pollution from metallurgical operations. Industries are investing in pollution control measures such as scrubbers, filters, and wastewater treatment systems to minimize the release of harmful pollutants into the environment.

3. Energy Conservation: Energy-intensive metallurgical processes are being optimized to reduce energy consumption and greenhouse gas emissions. Efforts are being made to enhance energy efficiency through the use of advanced technologies, waste heat recovery systems, and renewable energy sources. The integration of energy management systems and the promotion of energy-saving practices contribute to sustainable metallurgy in India.

4. Waste Management: Effective waste management is crucial for sustainable metallurgical practices. India emphasizes the proper handling, treatment, and disposal of metallurgical waste to minimize environmental impacts. Recycling and reprocessing of waste materials are encouraged to recover valuable resources and reduce the reliance on virgin raw materials.

5. Life Cycle Assessment: Sustainable metallurgy in India involves conducting comprehensive life cycle assessments to evaluate the environmental impact of metallurgical processes from cradle to grave. This helps identify areas for improvement and make informed decisions regarding materials selection, process optimization, and waste management strategies.

6. Research and Development: India's metallurgical research institutes and industry collaborations are actively engaged in developing innovative technologies and processes that align with sustainable practices. Research focuses on the development of cleaner and greener metallurgical routes, the use of alternative materials, and the reduction of environmental footprints.

7. Environmental Certifications: Metallurgical industries in India are increasingly adopting environmental management systems and pursuing environmental certifications such as ISO 14001. These certifications demonstrate a commitment to environmental stewardship and sustainability and help drive continuous improvement in environmental performance.

8. Stakeholder Engagement: Sustainable metallurgy in India involves collaboration among industry stakeholders, government bodies, research institutes, and local communities. Stakeholder engagement promotes knowledge sharing, the exchange of best practices, and the development of common sustainability goals.

Through these efforts, India is striving to make its metallurgical industry more environmentally friendly and sustainable. By integrating green practices, optimizing processes, and promoting responsible resource management, the country aims to reduce its environmental footprint while supporting the growth and development of the metallurgical sector.

Metallurgy for High-Value Industries

Metallurgy plays a crucial role in supporting high-value industries in India by providing materials, technologies, and expertise for their operations. Here are some key sectors where metallurgy plays a

205

significant role in India's high-value industries:

1. **Aerospace and Defense:** Metallurgical advancements are instrumental in the aerospace and defense sectors. Materials with high strength-to-weight ratios, excellent corrosion resistance, and superior mechanical properties are essential for aircraft, satellites, missiles, and defense equipment. Metallurgical research and development focus on developing advanced alloys, composite materials, and surface treatments to meet the stringent requirements of these industries.

2. **Automotive:** The automotive industry demands lightweight materials with exceptional strength and durability. Metallurgical advancements enable the development of high-strength steels, aluminum alloys, and advanced composites for automotive components. Metallurgists work on optimizing manufacturing processes, heat treatment techniques, and joining methods to enhance the performance, safety, and fuel efficiency of vehicles.

3. **Energy and Power:** Metallurgy plays a critical role in the energy and power sectors, including conventional and renewable energy production. Power plants, including thermal, nuclear, and hydroelectric facilities, rely on specialized metallurgical materials for boiler tubes, turbine blades, heat exchangers, and other critical components. Metallurgical innovations focus on improving material performance, corrosion resistance, and thermal properties to enhance energy efficiency and operational reliability.

4. **Electronics and Semiconductors:** The electronics industry demands high-purity metals, alloys, and thin films for integrated circuits, microelectronics, and semiconductor devices. Metallurgy plays a significant role in the fabrication and processing of materials used in electronics, including copper, aluminum, titanium, and specialty alloys. Metallurgical advancements enable the production of materials with precise electrical, thermal, and mechanical properties required for modern electronic devices.

5. **Medical and Healthcare:** Metallurgy plays a crucial role in the medical and healthcare sectors, particularly in the manufacturing of medical devices and implants. Biocompatible materials such as stainless

steel, titanium alloys, and shape memory alloys are used in orthopedic implants, dental implants, surgical instruments, and medical equipment. Metallurgical expertise ensures the production of materials with the desired properties for safe and effective medical applications.

6. Infrastructure and Construction: Metallurgy contributes to the development of robust and durable infrastructure and construction projects. Structural steels, reinforced concrete, and other metallurgical materials are essential for bridges, buildings, pipelines, and transportation systems. Metallurgical research focuses on improving the performance and longevity of materials used in construction, considering factors such as corrosion resistance, fatigue strength, and environmental sustainability.

7. Electronics Packaging and Interconnects: Metallurgy plays a critical role in the packaging and interconnects of electronic components. The development of microelectronic devices, printed circuit boards, and electronic packaging relies on advanced metallurgical techniques such as thin film deposition, soldering, and bonding. Metallurgical innovations enable the miniaturization, reliability, and functionality of electronic systems.

These are just a few examples of high-value industries in India where metallurgy plays a significant role. The metallurgical expertise, materials development, and technological advancements in these sectors contribute to the growth and competitiveness of India's industries, fostering innovation, sustainability, and economic development.

Skill Development and Education in Metallurgy

Skill development and education play a vital role in shaping the metallurgical industry in India. To meet the evolving needs of the industry and ensure a skilled workforce, several initiatives have been undertaken to promote skill development and provide quality education in metallurgy. Here are some key aspects related to skill development and education in metallurgy in India:

1. Academic Programs: Various universities, colleges, and institutes across India offer undergraduate, postgraduate, and doctoral programs in metallurgical engineering and materials science. These programs provide a comprehensive understanding of metallurgical principles, processes, and applications. The curriculum typically includes subjects such as physical metallurgy, extractive metallurgy, mechanical metallurgy, materials characterization, and industrial training.

2. Research and Development: Indian educational institutions and research organizations actively engage in metallurgical research and development. This involves conducting cutting-edge research, exploring new materials and processes, and collaborating with industries to address industry-specific challenges. Research initiatives focus on areas such as alloy development, advanced materials, process optimization, and sustainable metallurgy.

3. Industry-Academia Collaboration: Collaboration between industry and academia is crucial for bridging the gap between theoretical knowledge and practical applications. Many metallurgical institutions in India have established partnerships with industries to facilitate internships, industrial training programs, and collaborative research projects. This collaboration allows students to gain hands-on experience and exposure to real-world metallurgical practices.

4. Skill Development Programs: Skill development programs aim to enhance the practical skills and employability of metallurgical graduates. These programs provide training in specialized areas such as metal casting, welding, heat treatment, non-destructive testing, quality control, and materials characterization techniques. Skill development initiatives are often supported by government schemes, industry associations, and vocational training institutes.

5. Certification Courses: Several certification courses are available to professionals and students seeking to enhance their knowledge and skills in specific areas of metallurgy. These courses focus on specialized topics such as corrosion engineering, surface engineering, metallurgical failure analysis, welding technology, and foundry technology. Certification programs provide valuable expertise and recognition in niche areas of metallurgy.

6. Professional Societies and Associations: Metallurgical professionals in India can benefit from membership in professional societies and associations. These organizations, such as the Indian Institute of Metals (IIM) and the Materials Research Society of India (MRSI), provide platforms for networking, knowledge sharing, and professional development. They organize conferences, workshops, and technical events to promote collaboration and disseminate the latest advancements in metallurgy.

7. Government Initiatives: The Indian government has launched various initiatives to promote skill development and education in metallurgy. These include skill development schemes like the National Skill Development Mission and the Skill India Mission, which aim to provide training and enhance employability. Additionally, the government has established Centers of Excellence and research institutes to support advanced research and development in metallurgy.

Efforts in skill development and education in metallurgy in India are geared towards creating a skilled workforce capable of meeting the industry's demands. Continuous updates to curriculum, industry-academia collaborations, research initiatives, and skill enhancement programs contribute to the growth and advancement of metallurgical education and skill development in the country.

Government Initiatives and Policies

The Indian government has implemented several initiatives and policies to promote the metallurgical sector and support its growth. These initiatives aim to enhance the competitiveness of the industry, attract investments, foster innovation, and ensure sustainable development. Here are some notable government initiatives and policies in India:

1. National Steel Policy: The National Steel Policy provides a comprehensive roadmap for the growth and development of the steel industry in India. It focuses on enhancing domestic steel production, increasing steel consumption in various sectors, promoting research and development, and ensuring sustainable development of the sector.

2. Make in India: The Make in India initiative aims to transform India into a global manufacturing hub by attracting investments, promoting ease of doing business, and facilitating the growth of domestic industries. It has played a significant role in encouraging investment in the metallurgical sector and promoting the development of manufacturing capabilities.

3. National Metallurgical Policy: The government has proposed a National Metallurgical Policy to address the challenges faced by the metallurgical sector and promote its growth. This policy aims to strengthen the industry's competitiveness, enhance research and development activities, encourage technology adoption, and promote sustainable practices.

4. Research and Development Support: The government provides support for research and development activities in the metallurgical sector through various schemes and programs. These initiatives aim to foster innovation, promote technology development, and enhance the sector's competitiveness.

5. Infrastructure Development: The government has prioritized infrastructure development, including the establishment of industrial corridors, dedicated freight corridors, and ports, to facilitate the growth of the metallurgical sector. These infrastructure projects help in the efficient transportation of raw materials, finished goods, and facilitate the establishment of manufacturing facilities.

6. Skill Development Programs: The government has launched skill development programs to address the skill gap in the metallurgical sector and enhance the employability of the workforce. These programs aim to provide specialized training and upskilling opportunities to enhance the industry's human resource capabilities.

7. Export Promotion Schemes: The government has implemented various export promotion schemes, such as the Merchandise Exports from India Scheme (MEIS), to boost exports from the metallurgical sector. These schemes provide financial incentives and support to exporters, encouraging them to explore international markets and increase their competitiveness.

8. Ease of Doing Business Reforms: The government has

undertaken significant reforms to improve the ease of doing business in India. These reforms aim to simplify regulatory processes, streamline approvals, and create a favorable business environment for the metallurgical sector.

9. Environmental Regulations and Compliance: The government has established stringent environmental regulations and standards for the metallurgical sector to ensure sustainable practices and minimize environmental impacts. Compliance with these regulations is enforced through regular monitoring and strict enforcement mechanisms.

10. Foreign Direct Investment (FDI) Policies: The government has liberalized FDI policies in the metallurgical sector, allowing greater foreign investment and collaboration. This encourages technology transfer, promotes investments in advanced metallurgical processes, and supports the growth of the sector.

These government initiatives and policies demonstrate a commitment to promoting the growth, competitiveness, and sustainability of the metallurgical sector in India. By creating an enabling environment, encouraging innovation, and addressing key challenges, the government aims to position India as a global leader in the metallurgical industry.

Global Collaborations and Competitiveness

Global collaborations play a crucial role in enhancing the competitiveness of the metallurgy sector in India. Collaboration with international partners brings access to advanced technologies, expertise, and global markets, fostering innovation and growth. Here are some key aspects of global collaborations and the competitiveness of India in the metallurgy sector:

1. Technology Transfer: Collaborations with global partners enable the transfer of advanced metallurgical technologies to India. This helps in improving process efficiency, product quality, and overall competitiveness. Technology transfer initiatives can include joint research projects, licensing agreements, and knowledge-sharing programs.

2. Research and Development Collaborations: Partnering with international research institutions, universities, and industry experts facilitates collaborative research and development (R&D) efforts. Such collaborations promote knowledge exchange, enable access to cutting-edge research facilities, and foster innovation in metallurgical processes, materials, and applications.

3. Market Access and Global Supply Chains: Collaborations with global partners facilitate market access and integration into global supply chains. This enables Indian metallurgical companies to expand their customer base, explore new markets, and enhance their competitiveness on a global scale. Joint ventures, strategic partnerships, and international trade agreements play a vital role in this regard.

4. Skill Development and Knowledge Exchange: Collaborations with international institutions and experts provide opportunities for skill development, training, and knowledge exchange. This helps in developing a skilled workforce equipped with the latest knowledge and expertise, contributing to the overall competitiveness of the metallurgy sector in India.

5. Industry-Academia Collaboration: Partnerships between Indian metallurgical industry and international academia foster collaborative research, talent development, and technology transfer. Joint industry-academia initiatives enable the development of skilled professionals, promote research-driven innovation, and enhance the competitiveness of the sector.

6. Participation in International Conferences and Exhibitions: Indian metallurgical companies' active participation in international conferences, exhibitions, and trade fairs enhances their visibility and networking opportunities. It facilitates the exchange of ideas, promotes collaboration, and helps Indian companies stay updated with the latest trends and advancements in the global metallurgy sector.

7. Investment and Infrastructure Development: Collaborations with global investors and infrastructure development agencies contribute to the competitiveness of the metallurgy sector in India. Foreign direct investment (FDI) inflows enable the establishment of modern manufacturing facilities, adoption of advanced technologies, and expansion of production capacities, boosting competitiveness.

8. Policy Alignment and Regulatory Harmonization: Collaborative efforts with international stakeholders can help align policies, regulations, and standards in the metallurgy sector. This supports the harmonization of trade practices, quality standards, and environmental regulations, enhancing competitiveness and facilitating smoother international business operations.

India's growing participation in global collaborations, coupled with its strong metallurgy industry base, skilled workforce, and increasing investment in R&D, positions the country as a competitive player in the global metallurgy sector. By leveraging international collaborations and partnerships, India can strengthen its technological capabilities, expand its market reach, and achieve sustainable growth and competitiveness in the metallurgy sector.

Challenges and Opportunities

The metallurgical sector in India faces several challenges that impact its growth and competitiveness. At the same time, there are also significant market opportunities that can be tapped into. Here are some key challenges and market opportunities for the metallurgical sector in India:

Challenges

1. Raw Material Availability: The availability and consistent supply of high-quality raw materials, such as iron ore, coal, and other mineral resources, pose a challenge for the metallurgical sector. Ensuring a reliable supply chain is crucial for uninterrupted production.

2. Energy Costs and Efficiency: The high cost of energy and the need for energy-efficient processes are significant challenges faced by the metallurgical sector. Improving energy efficiency and exploring alternative energy sources can help reduce costs and enhance competitiveness.

3. Environmental Regulations: Stringent environmental regulations and compliance requirements pose challenges for the metallurgical sector. Balancing sustainable practices with efficient production processes is crucial to meet environmental standards and maintain public trust.

4. Skilled Workforce: The availability of a skilled workforce with expertise in metallurgy and related fields is essential for the sector's growth. Bridging the skill gap and fostering continuous learning and development are critical challenges that need to be addressed.

5. Technology Upgradation: Keeping pace with evolving technologies and adopting advanced processes and equipment is crucial for the competitiveness of the metallurgical sector. Access to the latest technologies and expertise is a challenge that requires collaboration and investments in R&D.

Market Opportunities

1. Infrastructure Development: India's focus on infrastructure development presents significant market opportunities for the metallurgical sector. Construction of highways, railways, airports, and smart cities requires a vast amount of steel and other metal products, creating a growing demand.

2. Automotive and Transportation: The automotive industry in India is experiencing robust growth, presenting opportunities for the metallurgical sector. Lightweight and high-strength steels, aluminum alloys, and other advanced materials are in demand for automotive manufacturing, driven by fuel efficiency and emission regulations.

3. Renewable Energy: The push for renewable energy sources, such as wind and solar power, creates opportunities for the metallurgical sector. Components like wind turbine towers, solar panels, and energy storage systems require specialized metallurgical products and expertise.

4. Defense and Aerospace: India's defense and aerospace sectors are expanding, driving the demand for specialized metallurgical products and technologies. High-performance alloys, titanium, and other advanced materials are required for defense equipment, aircraft, and space applications.

5. Export Potential: The global market for metallurgical products offers significant opportunities for Indian companies. With competitive manufacturing capabilities, quality products, and adherence to international standards, Indian metallurgical companies can tap into global demand and expand their export footprint.

6. Value-added Products: Diversification into value-added products, such as specialized alloys, precision components, and advanced materials, presents opportunities for the metallurgical sector. Meeting specific customer requirements and niche markets can help differentiate Indian companies and enhance profitability.

7. Recycling and Circular Economy: The increasing focus on sustainability and the circular economy presents opportunities for the metallurgical sector to develop recycling technologies and processes. Extracting value from scrap and promoting a closed-loop approach can reduce raw material dependence and contribute to sustainable practices.

Addressing the challenges and capitalizing on market opportunities require a collaborative approach involving industry, academia, government, and other stakeholders. Investments in R&D, skill development, technology adoption, and infrastructure development are key enablers for the growth and competitiveness of the metallurgical sector in India.

Conclusion

The chapter concludes by summarizing the key trends, challenges, and opportunities in the future of metallurgy in India. It emphasizes the need for sustainable practices, skill development, and global collaborations to drive growth in the sector. The chapter encourages continued investments in research, development, and infrastructure to unlock India's potential as a leading force in the global metallurgical industry.

www.ingramcontent.com/pod-product-compliance
Lightning Source LLC
Chambersburg PA
CBHW060829220526
45466CB00003B/1036